FAN SITES

ABBY S. WAYSDORF

FAN SITES

Film Tourism and Contemporary Fandom

University of Iowa Press, Iowa City

UNIVERSITY OF IOWA PRESS, IOWA CITY 52242

Copyright © 2021 by the University of Iowa Press
uipress.uiowa.edu
Printed in the United States of America

Design by Teresa Wingfield

Printed on acid-free paper

LIBRARY OF CONGRESS CATALOGING-IN-PUBLICATION DATA

Names: Waysdorf, Abby S., 1985– author.
Title: Fan Sites: Film Tourism and Contemporary Fandom / Abby S. Waysdorf.
Description: Iowa City: University of Iowa Press, [2021] | Series: Fandom   and Culture |
Includes bibliographical references and index. |
Identifiers: LCCN 2021003493 (print) | LCCN 2021003494 (ebook) |
    ISBN 9781609387921 (paperback) | ISBN 9781609387938 (ebook)
Subjects: LCSH: Tourism and motion pictures. | Motion picture audiences. |
    Fans (Persons)
Classification: LCC G155.A1 W39 2021 (print) | LCC G155.A1 (ebook) |
    DDC 338.4/791—dc23
LC record available at https://lccn.loc.gov/2021003493
LC ebook record available at https://lccn.loc.gov/2021003494

To Daniel Ralph

Contents

Acknowledgments

Books are always a process, and this one is no exception. It has been borne out of a long period of many late nights, frustration, confusion, and, ultimately, enlightenment (hopefully). I couldn't have done it without the support of many, many people, whom I will try to address here.

First, I would like to thank Stijn Reijnders, under whose supervision this book began as a PhD dissertation. I wouldn't have been able to do this research without him taking a chance on me, and I value his continuing support greatly. Without his feedback, this would have been a much lesser book. Thank you so much, Stijn. I would also like to thank Eggo Müller for his belief in my academic talents from when I first showed up in his classroom ten years ago to now, and whom I am privileged to have guiding me in my current work in Utrecht.

I am also grateful to my former research team at Erasmus University Rotterdam: Nicky van Es, who still needs to take me to a Feyenoord game; Balasz Boross, who always inspires me with his theoretical insight and wit (and thanks to Henrik for putting up with us constantly talking shop over dinner, as well as for your own friendship); and Leonieke Bolderman, without whom I'd never get anything done, who's been a rock, a shoulder to cry on, and a true inspiration as we fight through academia.

I have been very lucky in my career to be part of the fan studies community, who have encouraged and inspired me throughout my travels (both actual and figurative) with this work. Thank you to Rebecca Williams, Ross Garner, Mark Stewart, Kylie Stewart, Bertha Chin, Mark Duffett, Matt Hills, Simone Driessen, Abigail de Kosnik, Lori Morimoto, Paul Booth, and everyone else I've so appreciated spending time with at the Fan Studies Network conferences. You've continually reminded me why I love this work and why it's worth doing. I would also like to thank the editorial team at University of Iowa Press for their support throughout the process of making this book a reality.

I wouldn't have been able to do anything, though, without my friends who made the Netherlands a home for me when I came here ten years ago, and without whom being here right now would be so much more difficult. Thank you, Roel, Cate, Alwin, Pepijn, Tom, Ben, Noah, Morgan, Jonas, Derek, Courtney, Alexa, Linda, Marigje, Martijn, Felix, and everyone else for every dinner, late night, movie screening, flower market visit, and everything else that has made me feel welcome and loved. A special acknowledgment to my paranymph, Sam, who has always been there for me, never letting me get too low, and reminding me about what's important (and that travel is for fun). I love you all.

Finally, I want to thank my family. I may be far away, but I think of you all the time, and without you there is no way I would be where I am today. Thanks to my sister, Nina, whom I've been privileged to watch grow into something incredible. Thanks to my brother, Matt, who's become someone I constantly brag about, and his wife, Meaghan, whom I love having in the family, and to my nephew, Daniel, whom I haven't met yet but love already. And, finally, but most importantly, to my parents, whose love, support, and guidance encouraged me to follow my dreams, even though it took me far away from them. They have always been my biggest fans. I hope I've made you proud.

An earlier version of chapter 2 has been previously published as "The Role of Imagination in the Film Tourist Experience: The Case of *Game of Thrones*" in *Participations* 14, no. 1 (2017): 170–191; an earlier version of chapter 3 has been previously published as "Fan Homecoming: Analyzing the Role of Place in Long-Term Fandom of *The Prisoner*" in *Popular Communication* 17, no. 1 (2019): 50–65; and an earlier version of chapter 4 has been previously published as "Immersion, Authenticity, and the Theme Park as Social Space: Experiencing the Wizarding World of Harry Potter" in *International Journal of Cultural Studies* 21, no. 2 (2018): 173–188 (all with Stijn Reijnders). My thanks to these journals, their editors, and the reviewers for their valuable feedback, which greatly improved these texts, and for permission to reuse them here. Parts of the introduction were also previously published as "Placing Fandom: Reflections on Film Tourism" in *Locating Imagination in Popular Culture: Place, Tourism, and Belonging*, edited by Nicky van Es, Stijn Reijnders, Leonieke Bolderman, and Abby Waysdorf (Oxon: Routledge, 2021), 283–296.

FAN SITES

Introduction
Why Are We Here?

When explaining my research, I tend to get one of two responses. The first is excitement—people have stories about film tourism trips they've been on, or ones they want to go on, which they relay to me with enthusiasm or a little bit of envy. The second is confusion—why would anyone want to go see some field where they filmed a TV show? Isn't that a bit weird? Wouldn't it ruin the magic?

While exact numbers are hard to come by, the first group is not alone. Film tourism is seen as an important, and growing, niche in the tourist industry, with films and television shows serving to provide valuable advertising for destinations in an increasingly competitive tourist market. Having a popular film or television show showcasing an area's landmarks and landscapes is seen as not only a way to stand out from the crowd, but a way to draw tourists in directly to see where a favorite "took place." Stories about places that show increases in tourist numbers after their appearance on screen pop up regularly, from the remote Irish island of Skellig Michael, which appears in a recent Star Wars film,[1] to Taekbaek in South Korea, which plans to rebuild the torn-down film set of the hit drama *Descendants of the Sun* to satisfy the tourists who have already come looking.[2] Film tourism is a successful business and only looks to become more so.

Despite the confusion of the second group, in some ways the reasons for this practice are clear. From viewing great works of art and architecture to experiencing the customs and foods of faraway lands, the rhetoric of tourism tells us that there is no substitute for really "being there." It has long served to make the exotic and fantastic real, bringing what was previously only told about into the physical world of the tourist. Photography, film, and television have been integral to this process for over a century,[3] familiarizing us with what is worth experiencing in person. We feel as though we know places from seeing them in the movies or on television. Fictional texts can have this effect as much as nonfictional ones, giving us a sense of how people live in other lands in a more affective manner than nonfiction does, bringing us to *Sex and the City*'s New York City cocktail bars or *Amélie*'s Paris streets. This is not a new phenomenon—literature has inspired tourism for centuries,[4] becoming part of the official heritage discourse of many places. Tourism connected to film is therefore carrying on the long traditions of organized tourism, updated for an age in which film and television viewing are our favorite leisure activities.

Academics have taken notice of this trend, as we tend to do. Beeton,[5] Karpovich,[6] and Connell[7] have detailed the ways in which the academic field of film tourism has developed over the years, showing how "research has progressed from speculation through to justification, developing knowledge of the implications of the activity, and, finally, to refinement of methodological and theoretical approaches."[8] These studies range from tourism management perspectives and business and marketing research through to leisure studies and, increasingly, media studies, making a hybrid field for a very hybrid subject. Recent years have seen increased interest in the topic as well as a desire to bring these disparate threads together. Major edited volumes[9] and journal special issues[10] have put media and tourism researchers in dialogue to investigate how popular culture and tourism interact in the contemporary media environment, looking at the ways in which they shape each other. And all the while, new film-related locations and attractions rise to prominence, proliferating across the media and tourism landscape. Film tourism is no longer just something speculated about, or anecdotes traded among the marketing staff of tourism bodies. It is a real, recognizable phenomenon.

Yet, among all this attention, significant questions and gaps remain. Film tourism research, while varied and growing, is still mostly in the form of individual case studies, which provide insights into specific examples but remain somewhat isolated from each other, without an overarching theoretical framework to understand the practice as a whole.[11] Fandom and media researchers have tended to use film tourism as part of a bigger argument about fandom, spending only a small amount of time on it,[12] while the more comprehensive studies that exist, such as Beeton's *Film-Induced Tourism* or Roesch's *The Experiences of Film Location Tourists*, largely take a tourism or marketing management approach to the subject, which, while important, is only half the story.

Understanding film tourism also means developing a holistic understanding of the practice as a cultural phenomenon, and how it illustrates and complicates particular trends in the way we relate to media, narratives, and the concept of reality itself.

That is what this book aims to accomplish. I argue that film tourism can best be understood as related explicitly to fandom as well as tourist practice, merging the two perspectives. This carries on from recent research in film tourism from a media and cultural studies perspective, which has sought to understand how film tourism is experienced by fans and media consumers as well as its role in the media industry. I build on this turn by taking multiple fandoms, locations, and tourists into account in a deeper investigation of film tourism as a contemporary practice. Fandom and tourism both encourage particular ways of conceptualizing and playing with "reality," and I show how these ways are brought together and put into tension in film tourism in order to create a meaningful experience for the fan tourist. The popularity of this practice in a digital age, with fans willing to spend money and time on traveling to fandom-related places, has an impact on fan practices and the relationship between fans and the media industry. It is exploring these two aspects, the individual experience of film tourism and the broader fan practice around it, that makes up this book. By doing so, I present a new theoretical and empirical understanding of what film tourism is and can be.

What Is Film Tourism?

Simply put, film tourism "is tourist activity induced by the viewing of a moving image, and is accepted as encompassing film, television, pre-recorded products (e.g. video/DVD/Blu-Ray), and now extends to digital media."[13] This is a bit of an unruly definition, as much of what comes under the title of "film tourism" is not, strictly, film. Others have preferred the term "screen tourism,"[14] as it focuses on the way in which audiences view audiovisual material, rather than focusing on the medium specificities of television, cinema, online video, and so forth. Alternatively, some have preferred the term "film-induced tourism,"[15] which focuses more on the way in which film can act as a "pull factor" for tourism practices. Both of these alternative terminologies have their merits. I will, however, be using "film tourism" throughout this book, following Connell, who notes that "film tourism" is the term most adopted in the literature and industry. For the sake of clarity, connection to earlier research, and as befitting the more emic perspective of this research in general, it makes sense for me to use it.

However, this book also fits into a broader framework of "pop culture tourism"[16] or "media tourism,"[17] which is tourism connected to any form of media, such as literature, music, or even video games. Film tourism has its origins in the earlier forms of literary tourism[18] and has considerable similarities with music tourism as well.[19] Media are increasingly interrelated in the contemporary era, especially when it comes to their fan practices, and tourism is part of this. However, as a visual medium, as well as arguably the dominant form of media today, film and television have specificities that shape their related tourist practices. This book attempts to address the issues specific to film tourism, although in an age of "transmedia"[20] it is not necessary to draw the lines too sharply. To this end, I also use the term "fan tourism" to highlight the fannish aspects of what I am talking about here, and to note the similarities between film tourism and fan-inspired tourism connected to other media.

Prior research on the topic has, as Beeton,[21] Karpovich,[22] and Connell[23] discuss, been divided roughly into two camps: that coming from tourism management and studies, largely (although by no means exclusively) drawing on quantitative methodology and concerned with understanding the volume and motivations of film tourists-as-consumers and how to manage or create the demand, and that from media, cultural, and fandom studies, which largely uses qualitative research and is more concerned with the construction of meaning around these sites. It is in this latter camp that this book is ultimately situated. However, it is still important for those of us who study film tourism as part of media culture to understand it as part of tourism as well. I do that here mostly through cultural studies approaches to tourism. While I draw on some literature from tourism management, particularly the work of Beeton as one of the key figures in this field, my ultimate interest is in the cultural construction and experience of film tourism. Therefore, it is the qualitative, cultural side of tourism research that I make the most use of. Within this area, I am also more concerned about the tourist experience rather than tourism's impact. This means that I am less interested in representation of specific places on screen and the implication of this representation in terms of place narratives and management, a frequent focus of film tourism research of all kinds that I will mostly bypass here.

As Karpovich[24] discusses, the focus on meaning-making and representation in much of these qualitative studies of film tourism has led to a focus on trying to understand what the tourist is "experiencing"—the media or the reality. As with film tourism research more generally, this focus is roughly divided between a tourism studies approach and a media studies approach, which I work to merge here. The disjuncture between what is on screen and what actually is, and how this difference is interpreted, has proven to be an intriguing way to investigate what authenticity, a key issue in cultural tourism studies, means for tourists today. If, as MacCannell[25] argues in a seminal work, we become tourists in order to seek out "authentic" experiences denied to us in our everyday lives, what does it mean for tourists to increasingly seek out experiences that are entirely inspired by the media (and therefore potentially not "authentic" at all)? For media researchers, film tourism provides a way to investigate concerns about how we as audiences interact with and make sense of a pervasive media culture. This has been discussed in terms of power relations between audiences and industry, in terms of enforcing the industry's power through the construction and control of these spaces, or in terms of the experience of "entering into" the textual world through tourism and challenging the lines between imagination, the media, and reality. This book builds on this line of research, but with a stronger focus on the experience of the (fan) tourist and how this experience is incorporated into their practices and lives.

To do so, I focus on specific narratives, rather than on film/television as a whole in the form of "city location tours" or film studio tours. Connell's overview of film tourism research[26] shows that the term has been used to cover a variety of related activities, from visiting film festivals to film-related theme parks to places that inspire famous films (such as *Braveheart*'s Scotland, which was actually shot in Ireland) to actual filming locations, whether they be "out in the world" or in a controlled studio setting. The linking factor is the association with audiovisual texts, whether it be in a more general or a more specific sense. Thus, the idea of "film tourism" is very broad, and it is necessary to be specific within each study about what the main focus will be. The interview studies presented here—*Game of Thrones* tourism in Dubrovnik (a city in Croatia) and Northern Ireland in chapter 2, sixties cult *The Prisoner* and its main filming location of Portmeirion in North Wales in chapter 3, and the Harry Potter series at the Wizarding World of Harry Potter theme park in Orlando in chapter 4—deal with tourists who came to these locations for their association with a specific fantastic narrative, rather than a more general association with film(s). This puts the tourists I focus on within a subset of more general film tourism practice, as they are not just "influenced" by films, or interested in the media world more generally, but they make space in their holiday (or, in some cases, design their holiday) to deliberately visit film-associated locations. These tourists have a specific—fannish—connection to the narrative world, one that they further develop through tourism. The fourth case study and the subject of chapter 5, on "pop-up" immersive experiences, explores how this connection is promoted—and managed—by a media industry seeking more control over the practices of fans.[27] As this book shows, place is an important part of fandom, and control over its functions and symbolic value is a new frontier in the battle to shape fandom in the twenty-first century.

The Film Tourism Experience

This book is built around two main research questions, which surface in various ways throughout the chapters. The first is this question: how, and in what different ways, do film tourists experience places related to their object of fandom? This builds on prior research on the film tourist experience as well as on non-film fan tourism, with a larger focus on how fandom influences the tourist experience. In their respective chapters, I discuss the specifics of the tourist experience of each place. The fans involved in each case have their own way of experiencing and making sense of the places involved, utilizing their history with the text, their own interests, and their relationship with the fan community.

However, we can still see important commonalities. In chapter 2, investigating *Game of Thrones* tourism in Dubrovnik and Northern Ireland, I define film tourism as an "imaginative experience," building on Reijnders's[28] centering of imagination in the media tourist experience and McGinn's[29] concept of "imaginative seeing." An imaginative experience is an experience shaped by the imagination (in that it is tied as much to the fictional narrative as it is to what is actually perceived), which in turn influences the imagination as well. This reciprocity is crucial here, in that the experience of film tourism has as much potential to influence the imagination as the imagination does the experience.

I further divide the imaginative experience into three main "modes": hyperdiegetic, production, and historical. The hyperdiegetic mode, based on Hills's[30] concept of hyperdiegesis, is about experiencing the locations as their fantasy counterparts do, performing actions like the characters or imagining themselves in their world. The experience of tourism not only lets tourists feel as though they are part of the story but allows them to imagine the narrative world beyond the screen—what King's Landing would physically feel like, what it would be to joust. This mode is joined by the production mode, which is based on imagining the production of the show—how the producers found locations, got actors there, built the sets, and so forth. Fans not only are interested in imagining how the show came to be, but appreciate the work and effort that went into the process—for *Game of Thrones*, as a symbol of how much its broadcaster HBO invested in making it a quality production (especially compared with the negative way that fantasy is usually considered). Finally, fans are interested in the history of the places they visit, a historical mode of imaginative experience. *Game of Thrones* created a frame for understanding the "real" histories of Dubrovnik and Northern Ireland, making this history more interesting while also confirming some of the "historical accuracy" of the series.

These modes are one of the major developments I make here and can be seen, in some fashion, in each of the other chapters, but shifting in a way that indicates the differences between these locations and their fandoms. The imaginative experience at a location is therefore built on a sense of what "happened" in a location, where the imagination imparts a meaning onto an archway or rock that would otherwise be any other archway or rock, a meaning that can be interpreted in different ways. However, what is most important in any imaginative experience is that the fan is having the experience in their own body. The physical experience of being at the location enhances and shapes the fan's understanding of the text when they return to it, either via watching it again or even just thinking about it. It is this physicality that I ultimately come back to as my final answer to how film tourism is experienced. As I show in my elaboration of the imaginative modes, and demonstrated by the fact that I needed to come up with multiple modes even at one set of locations, it is impossible to come up with one totalizing answer to how these places are experienced imaginatively. But there is one commonality to all parts of the imaginative experience: the importance of the embodied experience of the location—of physically being there.

This kind of co-presence and embodiment is, as I discuss in chapter 1, considered a crucial part of the tourist experience. The multisensory, embodied experience of tourism is how we confirm the reality of a place. Film tourism serves to do this for films and television shows. When we visit places associated with them, as we visit other types of landmarks and locations, these films and television shows become more than screen fantasies—they become real. This has often been discussed in terms of an immersion fantasy[31] of putting oneself into the textual world, with embodiment as part of this fantasy. The textual world "leaks" into our own, becoming something that can be touched and moving beyond "mere" fantasy into our own reality.

As the following chapters show, however, immersion in the textual/filmic world itself is only part of the way in which fans, even the most devoted fans, experience film-related locations. The experience can be as much extratextual as it is textual. But all parts of the film tourism experience are embodied. Visiting these locations not only creates new knowledge about the text and its spaces but connects the fan personally to it. That they have been in the same place as the text, or in a place that allows them to be physically immersed in the text, creates a different connection to it than watching does. Fans often find the experience difficult to put into words, as it is entirely contained within the body. As it is a bodily experience, utilizing all the senses as well as the feeling of being in a place more generally, it is intensely personal, as nothing is more personal than one's own body. It is this embodied experience that allows the imaginative modes to function. The necessary physicality of tourism makes the imaginative aspects work, connecting the fan to the show or film in a way that the more detached experience of watching does not afford. Fan tourists experience their object of fandom at the most personal—the bodily—level.

What this suggests is that embodiment is as crucial to film tourism as it is to other forms of tourism, even if film tourists are visiting a place that doesn't actually exist. They experience the location as a personal connection to the film or television show that prompted their visit. It is not just the world of the show or film in the sense of its diegesis that they connect to, but the entirety of it—its production, background, the emotions of watching, and so forth. This separates them from nonfilm tourists, as the place is never entirely "itself," but experienced in relation to this external presence. It also means that the question of authenticity and the search for an "authentic experience," which is so prominent in cultural tourism studies, is different here. Film tourists are less concerned with whether these locations are an "authentic" representation of the cultures they are hosted in, or even if they are entirely like they are on screen. What matters is whether they can have this personal connection with the text, which can be established even if the place doesn't entirely look or feel like it does in the text (although, of course, being close helps this process considerably). Film tourism can even be seen as sort of nonrepresentational geography[32] but a different sort than is commonly meant by the term, as it is based on connecting to a particular form of representation rather than the everyday experience of place. However, the fact that film tourism is essentially embodied, and that the experience is based on the practices and performances of physicality, gives it more common ground than one might think. Explicitly foregrounding the embodied nature of film tourism is therefore necessary. It is not an incidental part of film tourism of less importance than the mental processes of imagining and meaning making; it forms the heart of the experience.

Film Tourism as Fan Practice

Individual experience is only part of film tourism. That brings me to the second question of this book: what role do film-related places play in contemporary fandom? Fandom, in terms of both the community of other fans and the experience of being a fan, is not an isolated moment. It is something that the fan brings to the location they visit, and something they carry with them once they leave. As I will discuss in the next chapter, fandom is an increasingly important part of people's lives, a way to form identity and make sense of the world around us. As film tourism has grown, becoming an established practice in tourism and fandom circles, it is worth asking how these places affect the practices of fandom today, both for fans themselves and for the media industry around them. The existence of places that fans can visit to connect to their fandom, and the visits that tourists make to these places, undoubtedly play a role in how fandom operates. It is that role that I consider here.

This work builds on studies of fan conventions, which discuss the importance of fans meeting in physical space, but these gatherings tend to take place in less emotionally resonant surroundings. Fan conventions are among the oldest fan traditions, dating to the 1930s, and have long provided a space for fans to meet each other and celebrate their shared fandom. Despite the movement of day-to-day fandom to online activities, fan conventions have thrived in recent years. The San Diego Comic-Con, for example, not only transforms the city into fan space for the length of the convention[33] but is an important part of the promotional cycle for major media products. Smaller, more focused conventions have become a booming business, with dedicated conventions for everything from traditional media fandom texts like *Supernatural* to Bravo reality celebrities. For the dedicated and well-heeled, there are even fandom cruises, which promise a higher level of intimacy with both fellow fans and show creators and actors.[34] The continued success of these events, and the growth of fan tourism, give credence to Zubernis and Larsen's suggestion that "the more 'virtual' fandom becomes, the more meaningful experiences in these physical spaces are."[35] We can also see this in the growth of exhibitions and museums of popular culture,[36] which memorialize the physical objects of a film or television show as important heritage to be preserved. Encounters with these objects are, like visiting film-related places, reminders that the object of fandom has a physical presence that leaves traces and is worthy of being remembered. As with tourism, physicality (still) matters.

Film-related places are therefore one of several ways that fans physicalize their fandom, but a way that has particular specificities that I will explore in this book. First and foremost, they locate the fandom. They give the fandom presence in the physical world, as much as they do for the film and television show itself. Fandom is largely thought of as free-floating, as films and television shows can be accessed from (almost) anywhere, but film-related places tie it to a specific place even more so than the temporary placemaking of conventions. This is not unique to film and television fandom: as Rodman,[37] Hills,[38] and Sandvoss[39] discuss, places like Graceland and stadiums play similar roles for music and sports fans, respectively, providing an anchor in the physical world for these fandoms. They "provide a form of permanence to what would otherwise be a potentially fleeting pre-verbal experience."[40] Film-related places, whether they are filming locations or created and adopted locations like the Wizarding World of Harry Potter and FriendsFest, do this for fans of films and television shows. In this, the fandom, in addition to the text, becomes more real, because it is given form in our world. It has a specific place and therefore the stability and groundedness that place provides.

Consequently, that the fandom has a place engenders specific effects. There is the experience of physical connection to the object of fandom. But these locations are also sites of commemoration—places where the fandom itself and the experience of being a fan, and not just the object of fandom, can be paid tribute to, similar to popular culture exhibitions and museums,[41] but with their own specificities. As I discuss in chapter 3, being at a filming location that is frequently visited by fans links the fan to the fandom's history. The fans that perform reenactments at the site of filming connect themselves not only to *The Prisoner*, but also to the fans that have performed these reenactments over the decades. They physically connect to their fandom's history and, in many cases, to their own memories. Similarly, the sheer volume of other fans at the Wizarding World of Harry Potter, a popular and crowded theme park, meant that the Harry Potter fans interviewed felt that they were physically surrounded by their fandom—they could feel the weight of the series's worldwide popularity. While this made the park crowded and the lines for attractions long, it also made them feel that they were not alone in their passion. The series clearly mattered to others, and standing there with these others gave them an embodied sense of this. Because of this established presence, a place of fandom provides a place to be a fan—to perform fandom and fan practices in a way that the fan might not feel able to do elsewhere.

Throughout the writing of this book I encountered many fans who don't act as a fan outside of these locations. While they might discuss their favorite show or film with family members and friends, they rarely venture into the online (and offline) spaces of discussion and community that are considered emblematic of contemporary fandom. They are not "participatory" in the same way that the more frequently studied fans are, which, as I discuss in chapter 5, can be capitalized on and directed in particular ways. There were many reasons given for this: lack of time in their regular lives, fear that these more "obsessive" fans will be less accepting, not wishing to be categorized as one of the "obsessives" themselves. However, it was clear that when on holiday in these places, they were more willing to perform fandom than they would be otherwise. The structures and rhetoric of being "on holiday" supports acting as a fan in a way that "everyday life" does not. The tourist has time to spend on more frivolous pursuits, such as fandom, and has the freedom to act as they might not do at home.[42] While in everyday life they might be afraid of being considered "dorky," or perhaps not "dorky" enough, while on holiday they can fully act like fans.

This is enforced by the places they visit. If these are places of fandom, they are the correct place to act as fans. Within these spaces, fans can do what would be inappropriate or strange elsewhere, such as recite dialogue or even reenact full scenes, wear fan-related clothing or accessories, take particular photographs, and so forth. These kinds of performances are perfectly appropriate, even expected, at places of fandom. Even the fans of *The Prisoner* I interviewed, who were often archetypal "participatory" fans, felt that some activities were far more appropriate in Portmeirion than elsewhere. While they might be able to gather, talk about *The Prisoner* and its themes, and make references to it outside of Portmeirion, it was in Portmeirion that these activities felt the most natural.

That they are the places of fandom also means that fans can meet each other there, as they might at a fan convention. This can be in a diffuse way, such as Harry Potter fans enjoying the feeling of being around so many other fans within the space of the Wizarding World of Harry Potter, or in a more concrete way, such as *Prisoner* fans building deep bonds with the other *Prisoner* fans they have come to know through visiting Portmeirion regularly over the years. However, there are differences. Being at a place of fandom, where they feel as though the space is theirs and where they can act fully as fans, is considered to be more special than meeting elsewhere. It also surrounds fans with this particular fandom specifically, rather than the increasingly multifandom spaces of fan conventions. For "nonparticipatory" fans, the space of the fan convention is also confusing and sometimes threatening in a way that places of fandom are not. Conventions have their own social norms and modes of behavior that nonparticipatory fans are unfamiliar and uncomfortable with. While they overlap, particularly for very popular fandoms, there is a difference in the role that they play in contemporary fandom. In many cases, the specificity of places of fandom, compared with convention spaces, also creates a sense of permanence. That fans of *The Prisoner* can return to Portmeirion, which has not changed significantly since its filming in the 1960s, and connect to both *The Prisoner* and their fandom of it there gives these fans a sense of ontological security.[43] The perceived timelessness of Portmeirion and its sustained connection with *The Prisoner* makes it feel like a "safe vault" of the show and its fandom, something that will sustain even if *The Prisoner* fades out of the general consciousness.

What this all suggests is that specific places are an important, but frequently overlooked, part of how fandoms operate. While not every fandom has a place—as *The Prisoner* fans were keen to point out—it is significant that some fandoms have one, and that it can be created, as we see with the Wizarding World of Harry Potter. The increase in film tourism and special events, particularly for objects of fandom that don't always fit into the fan convention space, like *Friends* as discussed in chapter 5, indicates that there is a desire among fans to have a place and to have that place fill this role, or at least that this desire can be created. At these spaces, we can observe the media industry's attempts to build a version of media fandom through place-based practices—including gathering, reenacting the text, and engaging in special consumptive practices.

There is every indication that in the future a specific place, and the roles it fulfills for fans, will be an important part of most fandoms. As Beattie,[44] Booth,[45] and Garner[46] indicate in their studies of the Doctor Who Experience in Cardiff, or Jones's discussion of *The Walking Dead* tourism in Atlanta,[47] it is equally likely that these spaces will be explicitly part of the media industry's management of fans—as a way to attempt to control them and the way in which they enact their fandom. Both of these developments mean that fandom researchers need to take place into consideration when studying contemporary media fandom. The way in which place is used and controlled is an important aspect of how audiences and the media industry relate to each other today.

Structuring Fan Sites

This book unfolds over five chapters. Chapter 1 sets up the theoretical argumentation of this book, arguing that film tourism can only be understood by understanding the discourses of both tourism and fandom studies. Contemporary tourism, as argued by Urry and Larsen[48] and Mac-Cannell,[49] is constructed around seeing important sights of elsewhere—the great works of art, spectacular landscapes, and so forth. The practice and the objects of the gaze are constantly circulated through different media, creating and reinforcing the idea that it is important to see and experience certain sites "for oneself." Visiting confirms their importance to us. It is not enough to see a picture: the tourist must be present with the object, see it with their own eyes, to truly experience it. Film tourism builds on this—"being there" is different than seeing a film on screen and can be incorporated into existing patterns and practices of tourism. It is also, potentially, more important on a personal and social level than visiting "traditional" tourist locations, connecting more strongly to the Romantic ideal of sightseeing.

This is because of fandom. Fandom is a way of finding one's place in the world in the contemporary age, both internally and, increasingly, externally through the formation of fan groups. The contemporary age has made this process almost synonymous with digital media, but as this research shows, physical place matters as well. Places provide a mode of connection. On one level, these places are sought out to give a sense of "reality" to what is only, if vividly, imagined. Visiting physical places tests and plays with the boundaries of what is real and what isn't, showing the differences between the two while allowing the pretense, even if just for a moment, that the boundary has collapsed. As I argue, though, the idea of "entering into" another world is only part of a larger issue of how film tourists and fans relate to the multiple worlds we regularly inhabit. This idea of using tourism to encounter the fictional and fantastic is now a standard practice, one that is recognized across fandoms. That it is so well-recognized suggests that it is enacted not only for its own sake, but because of what participating in it represents to the fan. Visiting these spaces not only plays with the interrelation between our world and that of the text, but provides a way to reflect on one's fandom. I suggest that visiting film-related locations serves to commemorate not just the text but being a fan of it—physicalizing the relationship that fans have with their fandom, memorializing what happened there and why it was important to them.

The first interview study, presented in chapter 2, is film tourism connected to *Game of Thrones* in Dubrovnik and Northern Ireland. Both prestigious and popular, *Game of Thrones* has been an unqualified success since its debut in 2011. Its mix of political drama, high-fantasy setting, and high production values have earned it a significant fan base, encompassing both "participatory" and less organized fans. The high budget provided by HBO means that it could do significant on-location filming in historical European landscapes, in keeping with its promise of being an "authentic" take on the pseudo-medieval high-fantasy genre. It is this tourism, on location in Dubrovnik and Northern Ireland, that is the focus of the first case. Conducted between the airing of the third and fourth seasons of the show, when Dubrovnik and Northern Ireland were the focal points of the show's tourism, this study is a glimpse at film tourism "in the moment," with a popular audience that is drawn to the "real" locations of filming as the show is airing. The newness of the series and its popularity meant that, in some ways, this can be considered "typical" film tourism, following the pattern of a successful film or television show that suddenly draws people to an area they might not have otherwise visited. This chapter investigates these fans' experiences at these places and the ways in which their imagination is engaged as they encounter the "real version" of what they have come to know on screen.

The second case addresses the issue of longevity in film tourism. One of the concerns about film tourism as a tourism driver has been questions about its sustainability.[50] It seems to appear very quickly and vanish almost as fast. Most studies, including the ones here, are focused on the present experience, rather than on a longer-term engagement with the place in question. *Game of Thrones* was visited "in the moment," and while many fans grew up with the Harry Potter series, the Wizarding World of Harry Potter as an attraction was fairly new at the time of research. However, it had been open long enough for some fans to develop a history with it, with some fans returning regularly and gaining familiarity with the space. As work in cultural geography suggests, repetition and return creates a different sense of, and attachment to, place than a single visit. That fans return suggests a very different use and role of place in their fandom than if they were only visiting once.

This line of inquiry is developed further in chapter 3, which analyzes *The Prisoner* and its fans' relationship with its main filming location of Portmeirion. Fans have been visiting Portmeirion since it was revealed as the show's filming location in 1968, with a fan convention taking place there near-continuously since 1977. This makes it an ideal case to investigate how a longer-term relationship with a place affects fandom, compared with the newness of *Game of Thrones* as both a fandom and a destination and the relatively recent creation of the Wizarding World of Harry Potter. That fans have been not only visiting but returning to Portmeirion for decades means that we require a new way of understanding how fans experience places and the role that visiting plays in their fandom. This case therefore introduces the concept of "fan homecoming," showing how the relationship with a film tourism location can develop over time.

In the third case, which is the focus of chapter 4, I analyze a different kind of film tourist experience. Film tourism research has focused on sites of filming,[51] with much of its findings based on the idea of connecting to the "authenticity" of these locations. However, it is not only filming locations that draw fans. Re-creations of film and television environments are increasingly popular, from small bars like Chicago's *Saved by the Bell*–themed pop-up restaurant[52] to large attractions like Comedy Central UK's touring FriendsFest (explored in chapter 5 of this book). This chapter deals with one such simulated environment, the Wizarding World of Harry Potter area of the Universal Studios Florida theme parks in Orlando. Built as a re-creation of several locations from the Harry Potter series as they appear in the films, this site promises fans immersion into the narrative world. The locations depicted—the village of Hogsmeade, home of the wizarding school Hogwarts, and the London neighborhood of Diagon Alley—are impossible to otherwise visit, in both a real sense (they are fictional) and a narrative sense (while set in Britain, they are forbidden to the nonmagical), but fans have long dreamed of the possibility of visiting them anyway. This case therefore looks at the way in which fans engage with simulated environments, and the way in which they respond to and make sense of such "unreality." I argue that the theme park, and perhaps themed environments in general, can be considered as a medium itself—and therefore the Wizarding World of Harry Potter is seen as an adaptation of the series into this medium, in line with the ideology of transmedia storytelling. How this adaptation is understood, engaged with, and utilized by fans is the focus of this case.

The final chapter builds on this development. As the interview studies show, place is important to fans and to fandom. It draws in fans who want more knowledge of and connection to their object of fandom and provides a physical location that grounds the feelings and practices of fandom. Place can also attract fans who are not normally "participatory" in the way that fan studies describes them, creating a space where they feel that fandom and all its practices are permitted and encouraged, all within the recognized frame of tourism as "outside of" real life. As chapter 4 shows, this does not need to be at the actual site of filming—if the adaptation is done well, fans will attach this feeling to a simulated place as well as to a "real" film location. Place therefore has a great deal of potential as a way to manage and market to fans, a new frontier in the struggle over meaning and control of popular texts.

This struggle is the focus of chapter 5, which looks at the marketing and promotion around official "immersive experiences" around the popular American sitcom *Friends*, looking at FriendsFest (friendsfest.co.uk), a touring festival now in its sixth year in the United Kingdom and branching out into Germany, Spain, Poland, Russia, and a UK Christmas edition, and the recent *Friends* pop-up in New York City. This chapter investigates how these experiences are marketed and promoted as places of fandom, what kind of fandom is encouraged in these spaces, and what this suggests about how the media industry sees the role of place in mobilizing audiences. I show how official places of fandom—like FriendsFest, but also the Wizarding World of Harry Potter—encourage a performative, but atomized, kind of fandom, centered around encouraging a relationship between the fan and certain iconic moments in the text rather than a relationship with other fans and potential transformations of the text.

Through this overarching framework and individual examples, this book demonstrates what place means to both individual fans and to larger fandoms. I explore how this works in different cases and different environments, and point toward the future of film tourism as a fan practice. It is to this exploration that we now turn.

The (Meaningful) Experience of Film Tourism

The actual act of visiting places because of their connection to a film or television show touches upon many intriguing debates. For tourism researchers, film tourism points to new directions in destination marketing research—what draws tourists to particular locations—as well as both challenging and enforcing long-held theories about how and why tourism, especially the practice of sightseeing, functions in contemporary society. For media researchers, it presents a glimpse into the way in which the borders between imagination, fantasy, and reality shift in the contemporary media environment, and showcases ways in which the media industry utilizes physical space in its power negotiations with the audience. Its interdisciplinary nature makes film tourism such an intriguing subject of study, but also a somewhat difficult one. The fields of tourism and media research do not always converge, and do not frequently consider the other when investigating the phenomenon of film tourism.

As a media researcher with a particular interest in the study of fans and fandom, I have personally approached the topic from that perspective. Those who make a point to visit places associated with a film or television show are, in some way, a fan of the text—they make an effort, whether it be large or small, to make a connection to it, and therefore tourism clearly falls into the category of fan practices. This interest in tourism, seen as a counterpoint to the digitized nature of contemporary fandom, is increasingly visible within academic circles as well as within fan circles themselves. However, understanding tourism itself is also integral to understanding film tourism and what kind of experience it is. Tourism research provides particular insight into the ways in which visiting places, particularly famous locations and sites, has become part of our standard repertoire of practices.

In this chapter, I explore film tourism theoretically, looking at how it can be conceptualized and framed, and consider different ways that this has been done in the past. I begin with an exploration of film tourism as a form of tourist practice, as this is how film tourism first rose to academic prominence, showing its connections to other tourist practices and particularly highlighting the importance of "co-presence"[1] and "embodiment"[2] to the experience of tourist locations. I also discuss the way in which film tourism both works with and complicates traditional concerns of tourism research, such as authenticity and the nature of sightseeing.

Following this, I explore film tourism as a media, and particularly a fan, practice, following on the work of Hills,[3] Sandvoss,[4] and Reijnders[5] on how film tourism links the fantasy world of the text to the physical world, and show how concepts from tourism studies can deepen this understanding. Of particular importance here is the role that the idea of "reality" continues to play in contemporary life, and the way this idea is constructed, valued, and ultimately made sense of by media audiences. This aligns fan studies with the so-called spatial turn in media and communication studies[6] in showing how issues of space and spatiality matter in a mediated, heavily digital environment. I will also consider the implications of the growth of the practice of film tourism for fandom, particularly in light of the emerging questions of power relations between the media industry and fans as the potential economic value of such visits—well documented by tourism researchers—becomes clear to the media industry. In this way, this chapter develops a theoretical understanding of film tourism as a practice and points to issues that will be explored in more depth in the case studies that make up the rest of the book.

Film Tourism as Tourism

As Connell states in her excellent overview of film tourism research, "as a research community, we are now aware that film tourism occurs, that it is part of a range of motivators in the tourism destination decision-making process, that it creates a range of impacts, and has been adopted by savvy tourism marketers and businesses seeking uniqueness and novelty."[7] Essentially, this means that film tourism is an accepted and well-recognized form of tourism, one that is increasingly prominent in tourism marketing as destinations attempt to differentiate themselves in an increasingly competitive market. There is clear acknowledgment within the field of tourism research, both sociological and management-driven, that film tourism is a practice worth considering. What is less clear is how, or even why, it is experienced by tourists. Connell herself calls the tourist experience "an emerging field of study"[8] within the broader research field.

This is not to say that there is no consideration of the tourist experience in existing literature. As Karpovich[9] discusses, the tourist experience has been approached differently by tourism researchers and media researchers. Tourism researchers are largely concerned with understanding film tourism as a tourist experience, either tying it directly to industry concerns of tourists' motivations, expectations, satisfaction levels, and specific activities at locations or linking it more theoretically to concerns of how place is represented and understood through these trips and what this means in terms of the "authentic experience" that tourists are thought to seek out. Media researchers, similarly, approach film tourism in order to consider its relation to media practices, investigating issues of media influence and power at these places and the interaction between reality and mediated fiction that these visits represent.

These studies show that film tourism is a multifaceted phenomenon, one that touches upon many aspects of contemporary culture and can be explored in a range of different ways. What's common among them, though, is that, for tourists, being at these locations is "a point of access to something 'special'":[10] a meaningful experience. By this, I mean that it is an experience that the tourist finds valuable and significant (emotionally, intellectually, or so forth), one that is worth the effort to have or even repeat. "Being there" at a filming location, while still a niche activity compared with visiting established tourist sites or even engaging in "film-induced tourism"[11] more broadly, is clearly something worthwhile for many. But what makes this so?

Part of the explanation can be found in the concept of tourism itself. Generally, tourism and travel are seen as an important part of contemporary life, a "secular ritual"[12] that is necessary to undertake for health, relaxation, and even education.[13] It is most frequently built around the practice of sightseeing, looking at notable things and places. When we go somewhere, we make a point of seeing its landmarks, taking in its cultural works, and "look[ing] at the environment with interest and curiosity."[14] For many, sightseeing is what tourism is all about, in both a positive and negative sense—it is "tourism's default [. . .] the only thing tourists are supposed to be good at."[15] Sightseeing is a basic building block of leisure travel as a cultural practice, a subject of parody and scorn[16] but also of acceptance and desire. To be a tourist is to want to see things outside of one's home.

Adler[17] traces the evolution of this visual focus of travel among Europe's elite, showing how the value given to traveling moved away from learning from and conversing with experts abroad as the fashion for "scientific" visual observation took hold. This was part of a general focus on the visual in Western society. The first "sightseers" thought of themselves as scientists, objectively observing and recording the landscapes and lives of others for an audience back home. As this market became saturated and the societal ideal of the neutral scientist was challenged by that of the Romantic aesthete, sightseeing became an emotional experience, "simultaneously a more effusively passionate activity and a more private one,"[18] that focused on the pleasure and enlightenment of the tourist as they gazed upon the extraordinary, alongside an increasing sense that it was "important" to see certain places and objects. The Romantic ideal also stressed the importance of "getting away" from everyday surroundings, particularly urban environments. Building on this, Urry and Larsen show how a combination of technological and social developments—railroads, workers' rights movements, educational norms, and so forth—moved sightseeing from exclusive to mass practice, making it "one of the characteristics of the 'modern' experience."[19] If it was valuable for the elites to sightsee, in both a moral and pleasurable sense, it was also so for the masses, who, after all, needed to be "educated." As it became possible for "everyone" to travel, it became assumed that everyone should—and that an important part of doing so was seeing the sights of elsewhere. In contemporary times, the practice of sightseeing is so unremarkable as to seem natural.

This, of course, does not mean that it is. As Urry discusses in his landmark concept of "the tourist gaze," looking—gazing—at places and people is a constructed practice, "conditioned by personal experiences and memories and framed by rules and styles, as well as by circulating images and texts of this and other places."[20] It is not a "natural" reaction to a site, but rather one that we learn how to do, that can be done differently in different circumstances. The "privileging of the eye"[21] in Western society meant that tourism became organized by the sense of sight—of looking—and what is worth looking at is determined by specific cultural values of the extraordinary, something "distinguish[ing] it from what is conventionally encountered in everyday life."[22] Ideas of beauty, strangeness, otherness, uniqueness, and so forth are enacted in the determining of what should be seen when we travel. MacCannell refers to this process as "sight sacralisation,"[23] the marking-off of certain attractions as important to view through labeling and promotion. This process traditionally appealed to the universal—the established great works of art and architecture, the spectacular natural views, the site of major historic events. In visiting them, we commemorate their importance, once again confirming that these are important places that need to be seen. As tourism has proliferated, however, more niche sites—such as filming locations—have also become worthy of the gaze.

The importance of promotion indicates the importance of the media to sightseeing. MacCannell called the "mechanical reproduction" of a site via postcards or newspaper articles "the most responsible for setting the tourist in motion on his journey,"[24] while Urry and Larsen state that the "anticipation of pleasures" that makes travel appealing "is constructed and sustained through a variety of non-tourist technologies, such as film, TV, literature, magazines, CDs, DVDs and videos, constructing and reinforcing the gaze."[25] Images of places, important landmarks, and travel are omnipresent in contemporary media culture, promoting both specific locations and the idea of travel and tourism itself. This is not a new phenomenon, as media have played important roles throughout the different historical stages of tourism, showcasing previously unknown destinations and creating appealing impressions of the experience of visiting.

These images then circulate widely, in advertisements, newspapers, magazines, television, film, and so forth, creating a shared cultural imaginary around what is worth seeing when one travels, and the value, both personal and social, of doing so. The role of media images in creating sights has led to some talk of a "hermeneutic circle"[26] or "circle of representation"[27] of tourism where "what is sought for in a holiday is a set of photographic images, which have already been seen in tour company brochures or on TV programmes,"[28] as viewing and photographing these sites for oneself "proves" that one has been to the location. Photography and film are considered important elements in creating and sustaining the tourist gaze by determining what is worth viewing and stimulating the desire to see it, and then to bring back proof that one has done so in the form of more media.

In recent years, tourist imagery is considered to have greatly proliferated, in line with the general proliferation of media in contemporary life. There has been an increase in travel programming on television and coverage in magazines and newspapers, alongside the "on-demand" media access of home video, the internet, and especially social media. As this type of programming first entered the culture, some predicted that it would lead to a decrease in the desire for corporeal travel as "seeing the sights" could now "be experienced in one's living room, at the flick of a switch; and it can be repeated time and time again."[29] Instead, "on-demand" tourist imagery ended up functioning in much the same way as older forms, increasing the demand for corporeal travel. As Jansson[30] and Månsson[31] show, the major change has instead been in who determines what the important sites are and how to view them, as tourists themselves join media professionals in showcasing their visits to the public. This has become a lucrative new mode of travel media.[32] Social media sites like Instagram encourage engagement and self-promotion through showcasing beauty and/or an enviable lifestyle, both of which travel pictures fulfill. Showing that you went somewhere special is a cornerstone of social media practice; there is now an entire new media ecosystem based around the visual promotion of travel by "influencers" and those who want to be them. Instead of a decrease in travel because of media, we instead see phenomena like film tourism—where the desire is to go to a place that has been vividly, and often frequently, seen.

If "seeing the sights" is so important, why is seeing them through photographs and films not enough? To answer this, we begin at another important aspect of Urry's analysis of tourism practice. He stresses that co-presence, "to be there oneself,"[33] is a crucial element of tourism. It is not enough to see pictures of a site—one must see it with one's own eyes for it to truly "count" as having been seen. Only through physical presence is an important site truly experienced, despite the potential of technology to re-create the visual encounter with increasing clarity. Tourism is a corporeal practice as much as it is a visual one.

Indeed, it is the importance of this corporeal element that is the most sustained critique of Urry's work. Tourism, after all, is not only sightseeing, even if that is arguably its most emblematic practice—it also encompasses a wide range of physical activities, from the exertions of "adventure tourism" to dancing at nightclubs to lying on the beach, and involves the movement and concerns of bodies at all points. Even sightseeing involves the presence of the body, and is understood through bodily action as much as the "meaning" of the site in question.[34] Tourism cannot be separated from the physical and temporal experience of being in a place—this experience is what defines it as a practice.

The concept of "embodiment" is particularly useful in exploring the necessary physicality of tourism. As Crouch argues, building on earlier work on "nonrepresentational" geography, we understand our environment through physical interaction with it. He argues that it is through embodiment—the "process of experiencing, making sense, knowing through practice as a sensual human subject in the world"[35]—that we understand a place. This understanding is based on practices of movement through space and engagement of all the senses, including but not limited to sight, and it is this physical encounter that ultimately gives a feeling of comprehension. As Rodaway[36] argues, sight might be the dominant sense in Western thought, but it is also the most commonly "tricked" through illusions and technology. Therefore, we must call upon other senses, which are (so far) less easily duplicated, to confirm the "reality" of something. This is not to say that there is no cultural or representational dimension to the meanings we give places, but that this dimension is only partial. Without the sensuous understanding gained through physical presence, it is felt that the knowledge of any location is incomplete.

If we take "embodiment" as a crucial factor of "knowing" a place, then we can see why "co-presence" with a sight is integral. To feel as if we have actually experienced it, we need to gain the embodied understanding that can only come with co-presence. Or at least, there is a lot of cultural belief in this idea. As Auslander[37] discusses in terms of theatrical and musical performance, there is considerable cultural and symbolic capital gained through being physically present at an event, even if technology can make a "better" experience (in the sense of comfort, visibility of the performers, cost, and so forth). This concept can be moved from live performance to live sightseeing easily. Even if the internet or a television show offers a better view of a particular sight, in terms of clarity and access, it is not culturally the same as "being there" with all its physicality and even flaws. Without embodied knowledge, it is seen as an incomplete experience. It is less "real."

With the ideals of co-presence and embodiment, film tourism as a practice comes more into focus. Being at a filming location is different than seeing it on screen, in terms of both how it looks and how it is experienced. For some, they are places that should be "seen for oneself" in the same way as more established tourist sites. Visiting them can be integrated into existing patterns of travel, both in terms of where filming locations are and the touristic practices involved. Film-related sightseeing involves much the same practices of gazing and performance as any other example of the form, and is increasingly part of industry practice as well, with tours regularly being offered and locations being marked out as sights.

Film-related sites become tourist sites for much the same reasons that other locations become tourist sites—they are seen as something out of the ordinary. In this case, what makes them extraordinary is what happened there, or what can be linked to the location. For some visitors, film-related sites offer advantages over established tourist sites. While certain places have long been culturally valued, this is often a general value rather than a personal one.

Tourists might feel obligated to see the Eiffel Tower or a battlefield, while not necessarily feeling the sort of personal connection that makes gazing on it a meaningful, emotional experience in the Romantic tradition. Sites connected to a favorite film or television show have this personal value—and often a strong social value as well, especially for others "in the know." Having been there is a potential source of "fan cultural capital"[38] that positions one as a knowledgeable and privileged member of the fan community.

However, in comparison to other tourism sites, film tourism sites are not only about understanding the location, but also about understanding the film or television show. Traditional issues of tourism representation and the tourist experience are experienced differently with film tourism, particularly the issue of authenticity. The location represents not only itself, but a fictional narrative, and it is experiencing the fictional narrative, rather than having an "authentic" experience with another place and culture, that is desired. While this is not to say that the identity of the "real" location plays no role, the fictional narrative takes precedence. Therefore, in understanding what makes film tourism a meaningful experience, we must understand the relationship that people have with these stories.

Film Tourism as Fan Practice

Why are film-related locations tourist sites? Beeton ultimately links this to a discourse of celebrity: these places are famous because of their involvement in film and television, and as they're famous, they're worth seeing.[39] This continues an earlier sightseeing tradition of "famous sites" but incorporates contemporary interest in the media and in celebrity culture. Because these sites have celebrity put upon them, they become of interest and are incorporated into the culture of sightseeing. This echoes the point made by Couldry in his landmark study of the *Coronation Street* set and Granada Television studio: that ultimately, these sites confirm the power of the media—that by showing a television studio or filming location as an extraordinary place, they signify that the world of the media is indeed more special than one's "ordinary" life.[40] The boundary between "real life" and "media life" is destabilized but confirmed to exist, with "ordinary people" and places on one side, and celebrities and the world of the media on the other. Ultimately, this also means that the importance of media itself, and the fame that it gives certain places and people, is the draw for these visits.

I do not want to suggest that these analyses are incorrect. Ultimately, fame and celebrity culture are of great importance in making film tourism exist, and it is crucial that we as scholars take a critical approach to the power structures that support these practices. However, I would also like to suggest that for many film tourists, it is not just proximity to fame, status, and the media that makes film tourism an appealing prospect—it is fandom.

Fandom has not been traditionally addressed within the literature on film tourism as a tourist practice.[41] While connections have been made between the experience of film tourists at locations and their emotional connections to the text, this is infrequently addressed in terms of fandom, at least not explicitly. While this is starting to change, as Roberson and Grady address, it is still a noticeable gap in the field.[42] This is curious because, while not every visitor to a film-related location is a fan, it is likely that fandom is, for many, at least part of the draw to such a place.

By this, I mean fandom as defined by Sandvoss: "the regular, emotionally involved consumption of a given popular narrative or text."[43] This is, as he discusses, a very broad, "lowest common denominator" definition of fandom, one that encompasses many who might not self-identify as "fans," but it is precisely this broadness that makes the definition useful. Other definitions of fandom in academic use since the first wave of "fan studies" in the early 1990s focus more on the classification of different types of fans, or define fandom more narrowly, which often creates a normative, hierarchical sense of what "real fans" do (such as form communities and produce their own media works) or are (resistant to dominant norms of society), compared with the less admirable "consumer." This comes out of fan studies' early base in cultural studies and its desire to "de-pathologize" the figure of the obsessive and deviant "loner" fan common in psychology and communication studies. Fan studies has shown how fandom can be an "indicator species" of the media ecology, pioneering practices and demonstrating certain relationships with the media industry that highlight trends and capabilities that can then be sold to a more general audience (see also chapter 5). In popular discourse, however, the "fan" is still frequently seen as deviant, someone whose devotion to a favorite text or object is excessive, or, in a more positive light, part of the "subculture(s)" of fan communities, where they form an alternative community that uses and transforms popular culture to their own ends.

As Sandvoss discusses, as does Hills,[44] while a definition of fandom that focuses on its productive and resistant subcultures has been useful in the project of studying fandom, it is fairly limiting as a defining factor of what fandom, in the sense of being a fan, is. Subcultural fandom shows the potential of fandom and fan practice as a way of reclaiming texts and stories from the media industry, but it is not the only, or even the main, way that most people connect to texts. By focusing on fandom's affective qualities instead,[45] we can use the term in a way that cuts across different fan subcultures while also allowing for those who are not "participatory" to be involved. This is not to say that the "exemplary" practices of organized fan communities[46] are not of interest (and indeed, the role of the organized and/or named fan community in film tourism reoccurs in each of the case studies present here), but that they are one facet of a broader cultural phenomenon—that of the regular, emotional engagement with texts. While there are certainly differences between fans in terms of emotional involvement and type of activity, the broader definition is useful in determining commonalities. All fans are emotionally involved with their object of fandom.

Arguably, not every example of film tourism is fannish. General film location tours, whether of a particular city[47] or of a production space,[48] are less fan-oriented, as their focus is to show filming sites for many texts rather than one specifically. However, when tourists deliberately visit a particular place because of a film or television show, as they do here, it is usually[49] because of a fannish attachment to a particular text. It is this attachment and emotional connection that drives them to make the effort to visit places connected to the text. As with many fannish pursuits, this can be arranged on a spectrum of effort, labor, and motivation, but the core element—the emotional connection to the text—is the common element, whether the fan is stopping by an easily accessible attraction on a family holiday or traveling to a faraway land with the intent of finding an obscure location.

This is because objects of fandom matter to fans. They have an affective power[50] that not only gives pleasure, but in some way shapes one's identity. They speak to something within the fan that can then be used to construct a sense of who they are as a person—what they value, what emotionally affects them, what helps them understand the world and their place in it. Hills suggests that fan objects function as a "transitional object," something that is both external and internal to the self. It is part of the fan's thoughts, helping to make up their sense of self, but also understood as originating and existing outside of them. There is some sense of ownership over it—as a special thing that the fan cares about and feels possessive over—that the fan can still share with others. These objects help fans to "manage tensions between inner and outer worlds,"[51] stimulating the sense of imagination that gives a vibrant inner life while maintaining and understanding the difference between this and the real world.

This mixture of internal and external in the way in which fans interact with fan objects also speaks to the "reality" of fictional texts. As Jenkins[52] and Saler[53] argue, favorite fictional texts are in some way "real" to fans. Their places and people are ones that fans find themselves thinking about, or even "gossiping" about, in a way that they might discuss "actual" people and places. Whether to themselves or others, fans talk about their favorite texts as if they were real, populated by characters that they can speculate about just like someone they know, taking place in worlds that can be wondered about in the same way as a faraway place. As both Jenkins and Saler are both keen to stress, it is not that fans believe that the fictional worlds of their texts are actually real, only that they feel "as if" they are—they can occupy the same kind of mental space as "real" things.

There is, though, a clear understanding that fictional places are, in fact, fictional and wholly imagined. Yet, as Reijnders discusses, imagination and reality are not entirely separate—rather, they influence and play off each other.[54] Imagination becomes more vivid when it intersects with reality, while reality becomes more meaningful as it touches upon important imaginaries. As he states, as "imaginations and realities are interwoven, people feel the need to unravel them"[55] by probing at the areas where they connect. Fans especially often enjoy playing with and testing these boundaries, by bringing fictional worlds into ours through practices such as cosplay and collecting or by searching out places where the fictional and real worlds meet. As has been often discussed, fans come to locations looking for the "reality" of what is otherwise entirely imagined.

As tourism scholars have shown, tourism is a recognized way of confirming the reality of a place or object. Sandvoss states that visiting such places "creat[es] a relationship between an object of fandom and the self that goes beyond mere consumption and fantasy."[56] By going beyond "mere" consumption and fantasy, Sandvoss stresses the way in which physicality adds substance to what is otherwise ephemeral. While a text might be real in an emotional sense, and vividly imagined, it is still somewhat incomplete, as what we understand as ultimately "real" comes out of a multisensory encounter. In creating one, fans further push the boundaries between real and imagined.

This relationship between the imagined world and the physical world has been at the heart of much previous analysis on film tourism. As the place where the world of the text and our world meet, the encounter with the textual world itself is frequently put forward as the prevailing force of film tourism. Hills conceptualizes these encounters as "sustain[ing] cult fans' fantasies of 'entering' into the cult text,"[57] a point that Roesch echoes by referring to a "place insiderness" where the tourist "takes on the personality of the film characters and simulates what they must feel and experience in the scene,"[58] what he considered the highest level of the film tourist experience. Immersing oneself in the text is, in this viewpoint, considered to be the ultimate goal of film tourism, suggesting that meaning is created if the tourist can get close to this experience. A similar perspective is taken in studies that explain film tourism through the textual elements of film and television, such as Reijnders's linking of James Bond tourism practices to the series's performances of masculinity[59] or Tzanelli and Yar's reading of *Breaking Bad* tourism as a chance for the "normal" tourist to "transgress" through imaginary encounters with the drug trade.[60] Here, as with Hills and Roesch, it is through entering or reenacting the text that the practice is ultimately given meaning.

While I do not want to suggest that the text itself, and the desire to "enter into" it, plays no role in the motivations and experiences of film tourists, I would like to add to this textual focus. Film tourism is, after all, connected to a wide range of texts with a wide range of stories, and the *Breaking Bad* tourist might act fairly similarly to a *Lord of the Rings* one, despite their strikingly different narrative themes and landscapes. An alternative perspective can be found in the work of Couldry. He argues that the *Coronation Street* set "functions as a material form for commemorating the practice of viewing television"[61]—that, essentially, going there commemorates the long existence of *Coronation Street*, and of the practice of watching television, which has undoubtedly been an important part of the tourist's life. A similar point is made by Reijnders in his concept of "places of the imagination,"[62] which provide "reference points" to imaginary worlds much as historical landmarks do for historical events, and by Hills, who notes that "cult geography" shifts the audience-text relationship "towards the monumentality and groundedness of physical locations."[63] These places are not just sites of entering into the text—they are places of commemorating the text.

To this, I would also like to add that these places commemorate not only the text, but its fandom. It is not simply that a text happened that makes it worthy of commemoration, but that watching it meant something to its fans. Williams refers to the relationship between fan and text as a "pure relationship,"[64] a form of relationship that is entered into only for the satisfaction it can deliver, and continued only so long as it continues to deliver this satisfaction. This relationship does not have to be reciprocated, as the fan enters into it for what they personally get out of the text or from other fans. For fans, being a fan of a certain object grants a sense of comfort, pleasure, self, and/or ontological security, and can therefore be very important in the fan's life and identity.

Commemorating this relationship is therefore also an important part of what makes film and television tourism (and perhaps other types of media tourism as well) a meaningful experience. As Hills discusses, one of the main functions of cult geography is that it "extend[s] the productivity of [the fan's] affective relationship with the original text, reinscribing this attachment within a different domain (that of physical space)."[65] In being at the place where "it happened" or somewhere that is otherwise designed to commemorate the text, the fan can celebrate their "pure relationship" with it. They can memorialize what happened there, that it was meaningful to them, and potentially even celebrate it with others. It makes the relationship with the text physically real as much as it makes the text physically real. Therefore, even if a place is portrayed negatively or is not particularly beautiful, such as the meth-riddled suburban Albuquerque of *Breaking Bad*, it can still attract film tourists. The desire to commemorate the enjoyable relationship of watching *Breaking Bad* is still present, alongside the desire to probe the boundaries of reality, media, and imagination. They are both part of the affective power and ultimate meaning of film tourism.

It is this affective relationship that leads to many of these trips being referred to, colloquially, as "pilgrimages." As Beeton[66] and Buchmann, Moore, and Fisher[67] suggest, there are significant similarities in form between the film tourist trip and the traditional religious pilgrimage, in that the film tourists make an effort to visit sites that are connected to "sacred" values, such as fame or the values given to the text, and desire proximity to "what happened" there much as religious pilgrims desire to be in the presence of a miracle site. Yet, as discussed by Sandvoss and Reijnders, as well as in chapter 2, referring uncritically to these trips as "pilgrimages" confers a solemnity and otherworldliness to these visits that is not necessarily present. Sandvoss suggests that "rather than a communal search for a future place in another world, they are individual journeys seeking a sense of place in this world,"[68] while Reijnders stresses the "suspension of disbelief" inherent to the media tourist compared with the religious one—that it is about playing with what is real and what isn't, which is fundamentally against the nature of religious experience.[69] These visits are about the past and present of the individual, rather than the hope for a better afterlife, and about the curiosity of the imaginative. Therefore, while pilgrimage can be a useful metaphor, we must be careful, as with any other attempt to map religion onto fandom, of using it simply because we recognize the similarities.

Rather, film tourism is tourism. It is akin to pilgrimage as much as all tourism is akin to pilgrimage, which is to say, similar in some ways and different in others.[70] At its core it is about confirmation and commemoration, which is what tourism (and especially sightseeing) has long excelled at. The film tourist gets a sense of the "reality" of the otherwise "unreal" screen landscape through the practice of tourism, having an embodied and co-present encounter with the screen-depicted space, which has long been recognized in our society as establishing the "reality" of something otherwise existing only in media. Traditionally, in visiting important sites of history or culture, these sites' value is confirmed and commemorated. Film tourism does this for sites of popular culture and fandom.

Reality, Fantasy, and Film Tourism

As shown, previous studies on film tourism have highlighted the importance of the tension between reality and unreality in making the experience meaningful. Film tourism affords a unique way of playing with the borders between reality and unreality, one that draws upon multiple discourses about what is "real" that highlight particular aspects of its use in the contemporary media environment. What I do in this book is explore the different ways in which "reality" is constructed via place and popular culture today.

To understand this, we must first look at what is meant by "real." "Reality" and "real" are tricky, fluid terms, with many definitions and many antitheses. It is perhaps best understood as a discursive concept, rather than one with a definitive definition, which illustrates the different ways in which it is used by both media audiences and tourists. Using "reality" and "the real" this way means that there is no "true" real to betray or overshadow, but rather that the idea of what is real, and why is it important, is situationally and culturally determined. While reality's changing nature might be of concern for various reasons, I'm not going to argue here what they are. Rather, I attempt to reflect the complex way in which the term is used in contemporary society, and from there suggest why it is important, especially for film tourism. It can be understood in three main, albeit related categories: the real as physically present and opposed to the imagined, the real as true and opposed to the false or fictional, and the real as opposed to fantasy or the fantastic. The juxtaposition and combination of these different kinds of "real" is part of what gives film tourism its appeal.

The first way of using "real" is in terms of physicality—something that can be touched and grasped in a multisensory manner. It is through our senses that we determine what is or feels real, and what isn't or doesn't. In contemporary society, as Rodaway discusses, this is frequently ultimately confirmed through touch, as touch is not only less simulated through technology (compared with the visual or auditory senses) but is also considered the most intimate.[71] Feeling something physically is perceptually difficult to disregard in the way that one might disregard visual or audio stimuli, which are associated with the simulative and unreal practices of the media—we're used to the eye and ear being tricked, and therefore audiovisual media doesn't seem entirely real. Yet even if an object is created or otherwise fictional, if we can touch it, it physically exists in the world. As discussed earlier in this chapter, this idea of physicality as the ultimate arbiter of reality connects to the importance of embodiment in tourism—to be physically present with something gives it a particular meaning, because it feels more "real." This is in comparison to the object or place as imagined, existing only in the mind (or perhaps the eye) rather than perceived through all the senses. Potentially anything can be imagined, but only what is physically perceived is real. Yet, as Reijnders discusses, imagination and perceived reality are not entirely separate—rather, they influence and play off each other.[72] Imagination becomes more vivid when it intersects with reality, while reality becomes more meaningful as it touches upon important imaginaries.

This brings us to the next use of the term: the real as true, in opposition to false or fictional. Something might be "real" if it "really happened," in the sense that it is verifiable. Touch, as mentioned, is a major way of verifying that something exists, but, of course, it is not the only one. In this use of the term, *real* essentially means "not fake," not made by or for some external presence, and can also work as an analogue to "authentic"—a "real" battlefield where a confirmed historic event took place, a "real" native craft as opposed to one machine-produced or made for tourists.[73] If something is not real, it can be seen as a sort of deception, a deliberate fake intended to defraud or manipulate an unsuspecting public.

Alternatively, we can see this creation as an integral part of artistic expression and appreciation. The counterpart to "real" then is not so much fake as fictional—something imagined, created, and presented with the awareness that it is not real. Audiences know this going in, and therefore it is enjoyable rather than deceptive. As Ryan suggests, appreciating the work that goes into making this "feel real"—an actor successfully depicting grief, a story that absorbs you into its world—is, for most audiences, an important part of appreciating art.[74] Indeed, a great deal of Western art has been built around the idea of creating a believable illusion of reality, with considerable technological development going into the goal of making ever more convincing representations of it.[75] Returning to Jenkins and Saler, the fact that fiction often "feels real" makes it also feel meaningful. Emphasizing this is not without controversy, particularly from the more "highbrow" cultural critics who connect fictionality to fakery and warn about the "dangers" of getting too absorbed in fiction or technology and forgetting what is real. However, fictionality and illusion, in their playful but ultimately reciprocal relationship with reality, are undoubtedly a major part of Western cultural life.

The third use of "reality" is built on this one—namely, around the idea of reality as opposed to fantasy or the fantastic. In this use, "real" is considered more in terms of "realistic," or whether a piece of fiction or use of the imagination corresponds to reality as we know it. As with "real" itself, "fantasy" can be difficult to define, even if we just confine it to created works. As Attebery discusses in his attempt to define fantasy literature, it potentially encompasses a great deal—everything that does not exist in our reality can be put under the "mode" of the fantastic.[76] Fowkes, dealing with fantasy film, defines the genre as one in which "the audience must at the very least perceive an 'ontological rupture'—a break between what the audience agrees is 'reality' and the fantastic phenomena that define the narrative world."[77] Essentially, fantasy is "unreal" even beyond its core fictionality— it is something that does not or could not happen, such as magic, talking animals, interstellar starships, and so forth. It is tied even more directly to the imagination, as it relies on the mind's power to come up with what cannot be perceived at all, and requires a great deal of creativity to produce. In literary theory, this is opposed to "mimesis"[78] literature, which supposedly depicts the world as it is, in a realistic fashion. Yet, as Attebery discusses, even here the relationship between reality and fantasy can be seen more as a continuum rather than an absolute separation. Mimesis without fantasy is dull, while fantasy without mimesis is nonsensical.[79] Both must be related to the other in order to work properly, at least in a narrative context.

Within film and television, "the fantastic" is usually broken down into three main categories, science fiction, horror, and fantasy, the borders of which are somewhat blurry. Fowkes, for example, excludes science fiction and horror from her definition of fantasy, as they are said to be explorations of things that could exist, lacking the complete "ontological rupture" of fantasy.[80] Johnson, on the other hand, includes science fiction and horror in her discussion of "telefantasy," fantasy on television, because fans of such programs do so, which creates a genre by default. She links the programs of this genre together by their "non-verisimilitude," emphasizing that they share the same issues of "how to represent what 'doesn't exist.'"[81] In keeping with my general focus on fans and their use of terminology, I am following Johnson's definition. Here, fantasy is what does not and could not exist in the "real" world but is nonetheless depicted on screen, compared with a more mimetic fiction that theoretically could exist.

Film and television have a particular relationship to the fantastic. Compared with fantasy literature, which relies on the imagination of the author and reader, these places must be portrayed in a way that seems plausible to the eye and ear despite the viewer's knowing that they are fantasies. Audiovisual verisimilitude is integral to cinema and television, with their appeal largely built on how accurately they can reproduce the looks and sounds of "real life."[82] Fantasy manipulates this understanding of film, bringing what can't be to "life" in the sense of successfully convincing the viewer that it could exist.

For "live-action" (as in, not animated) film and television, there are physical practices involved in creating the fantastic world that is visible on screen.[83] Practices of filmmaking, set design, costuming, acting, and so forth are used to create the illusion of the fantastic, and give these narratives a connection to the "real" world through their very physicality. While the world is still unreal—as in fictional, as in fantastic—the actual creation of it leaves physical traces. In this, too, fantastic film and television is different from fantasy literature, in that literature does not leave such traces behind. While there might be places and objects involved, they are not so directly tied to what "happened" there, in terms of content production.

Compared with more mimetic audiovisual media, however, fantastic film and television has a different relationship between reality and unreality, especially in terms of place. Fans of a mimetic show or film gain a "sense" of the place where it is set via the way it is depicted on screen and might visit the place to see how closely the depiction does or does not match up. A fantastic place, however, only exists on the screen (or possibly the page). The fan tourist can only ever visit the representation, not the "real thing." Therefore, those visiting these places have, potentially, a different sense of what is important about the relationship between real and unreal. As such fantastic environments proliferate, becoming part of the transmedia strategy of the media industries that market such texts, understanding how fans interpret and make meaning from them is integral to understanding the way in which place, fantasy, and the media industry interact in the contemporary age.

The cases in this book deal with such fantastic audiovisual places: ones that do not exist in our reality, but are portrayed in film or television. Visited here are the high-fantasy Westeros of *Game of Thrones*, the mysterious Village of *The Prisoner*, the wizard-only London neighborhood of Diagon Alley and town of Hogsmeade from the Harry Potter series, and the "New York" of *Friends*. They differ in their subgenre of fantasy, histories, and fan relationships. This difference works to explore the range of issues associated with film and television tourism and, ultimately, the meaning-making processes of the fans who take part in it.

The first case, *Game of Thrones*, which is covered in chapter 2, is a television show based on a popular novel series in the "high fantasy" category, which is to say that it takes place entirely in a secondary world separate from our own.[84] As with much of the genre, this world is based on the culture and aesthetics of medieval Europe,[85] featuring kings, queens, castles, sword fighting, and so forth, combined with magic and mythical creatures like dragons. Medieval high fantasy has been a popular literary genre since J. R. R. Tolkien popularized it in the mid-twentieth century,[86] and *Game of Thrones* builds on this popularity and cultural familiarity. However, as I will discuss, it also positions itself as a "realistic" version of the genre, one that is closer to how medieval life "really was" compared with others. Part of its rhetoric of realism is based on its use of locations. The show is largely filmed "on location," utilizing historic landmarks throughout Europe to promote this sense of historical verisimilitude. Therefore, tourism connected to the show interacts with these locations and their "real" stories, theoretically unconnected to the televisual narrative imposed on top of them.

Comparatively, the Harry Potter series, the subject of chapter 4, is what is termed "low" fantasy, meaning that it takes place in "our" world. The lead character, Harry Potter himself, is raised in an ordinary English suburb until his eleventh year, at which point his heritage as a wizard, along with an entire parallel wizard culture, is revealed to him. The wizarding world exists in our own reality, hidden from those without magical talents or heritage—but can be revealed given proper permission. Many of its primary locations are part of "real" Great Britain, from the main setting of the Hogwarts school in Scotland to the London neighborhood of Diagon Alley. Rather than a completely other world, the Harry Potter series plays with the boundaries between reality and fantasy by suggesting that its environments could exist without any of us knowing it. Yet, of course, they don't. However, they can be visited not only at filming locations, but also at the Universal Studios theme parks, which have re-created several story locations as they look in the eight Harry Potter films. It is therefore in some ways a place of multiple unrealities—a re-creation of what does not exist, utilizing none of the "real world" locations that the series draws upon. In other ways, it is very real, physically existing in a way that most fantastic fiction does not. It therefore has very different answers about the nature and purpose of the relationship between fantasy and reality.

Portmeirion, the focus of chapter 3, is also somewhat of a re-creation, but in a very different way. A hotel complex and "home for fallen buildings" designed by Sir Clough Williams-Ellis, it combines a jumble of architectural elements and brightly painted Italianate features that set it apart from everything else in North Wales, and certainly it is far removed from an "authentic" Welsh village. Its fantastical appearance and isolated setting also made it an ideal filming location for *The Prisoner*, a television show about a spy (we think) who is kidnapped from his London home after his resignation and taken to a mysterious place known only as the Village. In the Village, he attempts to escape while being interrogated by a rotating cast of imprisoners as to the reason for his resignation. Drawing on a range of genres, from spy intrigue to science fiction, the narrative never fully reveals to the viewer what is going on—including exactly where the Village is, and what purpose it holds for whom. It blurs the lines between our reality and the show's fantasy even further by questioning what reality even is. What fans can hold on to is the equally fantastic Portmeirion, which they have been visiting for over fifty years now. It is not necessarily the Village, but compared with the historic locations used in *Game of Thrones*, it is not quite "real" either.

Finally, the fourth case, *Friends*, returns to our world, but a re-creation of it. While set in New York City, very much a part of the real world, *Friends* presented a fantasy version of it. This was facilitated by its being filmed entirely in Los Angeles, on a Warner Bros. sound stage. All the important locations of *Friends* are imagined—the building that houses the characters, the coffee shop they hang out in, and even their workplaces are all fictional, with no spot in the real New York City. While skyline shots feature in establishing that the story takes place there, that all the action actually takes place in sound stages, and that the characters rarely venture outside the established fictional locations, suggest that the show takes place in an idea of New York rather than the actual thing.[87] While not a fantasy in the way that *Game of Thrones* is a fantasy, it is not exactly true to life either. Rather, it shows the way in which media and storytelling can blur the boundaries between reality and fantasy, creating alternate, fantasy versions of places that can then be brought back into our real world.

Film tourism is a growing business, as well as one with a surprisingly long history. I suggest that it is the value of "reality," in each of the forms that I discussed, that is at the heart of this practice. Fans visit these places to access some form of "reality," in terms of real as physically present, real as nonfictional and/or authentic, or real as corresponding to reality as we know it (mimetically), for something that would otherwise be almost entirely unreal. The form of reality varies, as do the ways in which this search is enacted. This book investigates different ways in which the reality of fantasy is sought out and experienced, from filming locations to theme parks, from "new" destinations to those that have been at the heart of fan practice for decades. It is a multifaceted investigation of film tourism at this point in media and cultural history, and of the meaning that fans make out of encounters with "reality" today. Ultimately, I demonstrate that even in a heavily mediated age, experiencing something physically real has value.

Game of Thrones
The Role of Imagination
in the Film Tourist Experience

Perched atop a hillside, overlooking the medieval walled city and the brilliant blue water below, Fort Lovrijenac was originally built in the eleventh century to protect the city of Dubrovnik from seaside attack. Our guide tells us the story of its thick limestone walls and of the cannons that once stood there, but we are primarily here for another reason: Fort Lovrijenac was the filming location for many scenes of the Red Keep—castle of the royal family in the hit HBO television series *Game of Thrones*. Our tour visits the place where King Joffrey's name-day tournament was held and Cersei Lannister had her dramatic conversation with Petyr Baelish, where we stand on the old stones and peer at the binder full of screenshots to make sure we're in the right place as well.

It is not surprising that we are here. Film tourism is mainstream, with regions making filming sites a cornerstone of tourism marketing campaigns and newspapers and magazines offering guides to filming locations. In this chapter, I argue that considering the imagination is imperative to deepening our understanding of the film tourist experience. Strictly speaking, all travel is an imaginative act, but the role of the imagination becomes particularly prominent when we look at film tourism. As I argue here, the imagination plays a crucial role before, during, and after the concrete film tourist experience. Being there "in person" can enhance the imagining the fan already does, confront the fan with a sense of "reality," and stimulate the tourist's imagination after the return home. Imagination should be a core concept in studying film tourism. In this chapter I ask how, and in what ways, film tourists involve their imagination in practice when experiencing film locations.

What follows develops in two sections. First, I look at the role of imagination in tourism and fan practice more generally, developing the argument I built in the previous chapter. I then examine the specifics of *Game of Thrones* tourism in Dubrovnik and Northern Ireland, focusing on my field interviews, in order to investigate how film tourists' imaginative experiences occur in practice. From this, I identify three main modes of imaginative experience: the hyperdiegetic, the production, and the historical. Each of these contributes to the way that fans experience these locations, showing that the imaginative experience of a filming location is a multifaceted one, even for the same "group" of fans.

Imagination and Film Tourism

Imagination has always been a tricky subject—prone to mistrust and confusion, as well as delight and wonder. Theorists from Kant onward have tried to figure out and classify this peculiar human habit of picturing and thinking about what isn't directly in front of us. The American philosopher Colin McGinn, for example, stresses the generated nature of imagination in order to separate it from perception. According to McGinn, imagination requires human activity: "forming an image is something I do, while seeing is something that happens to me; in short, imagining is a mental action."[1] For McGinn, it is through this action that the world is made sense of, and given meaning, allowing us to contemplate what is possible outside of direct sensory input, offering an idea of what "could" be (whether it is physically possible or not). Imagination is therefore the mental activity of thinking about what is not directly perceived and envisioning what is possible.

The separation of imagination from perception seems to disconnect imagination from experience. Yet for tourism, and especially for film tourism, imagination and experience are highly connected. Rojek argues that "myth and fantasy play an unusually large role in the social construction of all travel and tourist sights."[2] We have already imagined a version of where we might go before visiting. Since our perception is based on their existing presence in our culture, tourist sites are already socially constructed as "extraordinary" places.[3] Hennig states that tourist activity is therefore based on widely held "myths" concerning themes such as freedom, nature, or paradise. It is through these myths that tourism gains meaning; as a practice it "takes place simultaneously in the realm of the imagination and that of the physical world."[4]

Film tourism goes beyond the general tourist imagination in that it connects place to specific fictions. Reijnders's concept of "places of the imagination" is a good starting point to understand how this works. He defines this as "material reference points like objects or places, which for certain groups within society serve as material-symbolic references to a common imaginary world."[5] For Reijnders, such locations allow a play with the boundaries between the real and the imagined, becoming spaces "where the symbolic difference between these two concepts is being (re-)constructed by those involved, based on what is considered 'factual' evidence."[6] Thus, visiting a film location is almost like visiting one's own imagination, but brought into real life. At the same time, the imagination is confronted with the physical reality of the specific environments, creating a powerful dialectic between imagination and perception, between symbolic landscapes and physical reality.

But if imagination and perception are different things, how can we resolve the imaginative nature of visiting film-related places? McGinn's notion of "imaginative seeing" ("seeing-as") is useful here: "the image comes to permeate the percept, to inhabit it, reach out to it, clothe it."[7] Essentially, in imaginative seeing, perception is given new meaning through imagination. It is "seen as" something beyond what is directly sensed. McGinn's definition, however, ignores how perception can also influence the imagination. Within film tourism, "seeing" each new place involves a process in which both parts influence each other. Imagining what "happened" in the fictional narrative when visiting a site gives it meaning, but physical experience of the actual place also influences imagination. This process is not necessarily strictly visual and can involve other sensory input. Rather than just "imaginative seeing," we can therefore describe film tourism as an "imaginative experience," an experience pre-shaped by a mental process of imagination and followed by a process of reshaping our fantasy lives.

This is complemented by the way in which fans are said to imagine and think about favorite texts. According to Matt Hills, most fan texts feature a form of "hyperdiegesis: the creation of a vast and detailed narrative space, only a fraction of which is ever directly seen or encountered within the text."[8] Their narrative worlds feel complex, inhabitable, and able to be explored. While watching *Game of Thrones* a viewer will see only a portion of its geography; fans often desire to complete this missing space. Tourism theoretically offers a way to explore hyperdiegetic spaces in a physical, embodied manner. Hills therefore refers to fans visiting filming locations as utilizing "cult geography," "extend[ing] the productivity of his or her affective relationship with the original text"[9] into physical space.

It is in this way that the relationship between fan and filming location has generally been theorized, as I talk about in chapter 1. Roesch builds the concept of "mental simulation" in his study of film location tourists: the tourist at a location disconnects from what is actually there and instead "takes on the personality of the film characters and simulates what they must feel and experience in the scene."[10] This is a fundamental component of a "place insiderness" where the tourist is "able to consume the imaginary beyond the filmic gaze"[11] and connect to it on what the tourist considers the highest level. In other words, location is theorized as helping fans imagine, explore, and insert themselves into a cult text's hyperdiegetic space, the space beyond the actual text, as tourists connect the physical space to the media object. As film and television fans are believed to spend time imagining what the world of their favorite narrative is like, expanding it beyond the screen, visiting the "actual" place is thought of as a way to make this imagining more concrete. Visiting their story's real location potentially gives them "embodied" knowledge,[12] moving their encounter beyond something merely cerebral and leading to a deeper experience of the story-world. Garner terms this potential "transmedia tourism," a kind of tourism designed to expand on a given story-world, with place used as a way to learn new things about it much in the way that other transmedia expansions are.[13] Therefore, fan tourists are seen as "actively [seeking] to re-vision the landscape and see beyond the physical geography"[14] into the story-world, which, as Lee discusses, expands the possibilities of the text and gives the tourist a sense that they can expand their knowledge of the narrative and its setting.

Yet, the grounding of these narratives in physical space also connects them to reality in a more prosaic sense—it is not only that fictional locations gain physical presence, but there are real locations involved, with their own stories, and real work involved in creating what is seen on screen. Not every fan focuses their attention on the diegetic world of a narrative—there are many ways to be a fan, and contemporary media texts have many elements to focus fandom on. Jones[15] suggests that some fans seek out locations not for the immersion, but instead for a sense of dissonance. For these fans, the difference between the actuality of the place and its screen counterpart is what makes film tourism interesting. In chapter 1 I discuss the importance of the physical practices of fantasy film and television—that in making these environments convincingly "trick" the eye into being perceived as real, there is work done in our world. This, too, is something that can be imagined, as a sort of fantasy career, one that is more interesting and glamorous than the fan's everyday life.[16]

The idea that dissonance matters as much as immersion to fans highlights the importance of a holistic understanding of how fans imaginatively experience places. The story-world is only part of the story, even for a transmedia property. The focus on the story-world as the most important guiding factor of film tourism has overshadowed other ways in which fans can have an imaginative experience with a place. As I will now explore through the way that *Game of Thrones* fans relate to and imaginatively engage with its filming locations, fans engage with favorite texts—and text-related places—in multiple ways, with both dissonant and immersive imaginative experiences.

Methods

The HBO television show *Game of Thrones* was a popular high-fantasy series based on George R. R. Martin's book series A Song of Ice and Fire. The series's sprawling story lines focus on several families and individuals vying for the Iron Throne of the Seven Kingdoms of Westeros; the major conflict is between the Lannisters in the capital city of King's Landing and the Starks of the North. As is typical of the genre, the show takes place in a world entirely outside of our own, but draws on the cultures, activities, and visual presentation of medieval Europe alongside recognizable fantastic elements. Therefore, there is no "real" version of King's Landing, but it clearly has historical referents. The production of the series is based in Northern Ireland, which also provides many locations for different parts of the Seven Kingdoms, especially the North, while Dubrovnik stands in for King's Landing and assorted locations in the outer lands of Essos. The show was highly successful in its eight-year run, with record-high viewership for HBO, an active fan base, and a prominent place in contemporary popular culture. These factors, along with the use of existing heritage locations, make *Game of Thrones* an ideal case study for examining the interaction of fandom, imagination, and the experience of place in film tourism.

In the summer and fall of 2013, between the airing of the third and fourth seasons of the show, fieldwork was conducted around four *Game of Thrones*–related sites and activities. While the show had also filmed in Iceland and Morocco, tourism had concentrated on Dubrovnik and Northern Ireland because of the number and accessibility of their filming locations. Interviewees were selected from tourists attending a commercial walking tour in Dubrovnik sold through a popular travel website, a recently launched commercial coach tour in Northern Ireland, and a fan-run coach tour that was part of an annual nonprofit Northern Ireland fantasy convention called TitanCon, and individual tourists at Fort Lovrijenac in Dubrovnik—a location prominently featured in guides to the show and easily accessible from the city center. A follow-up visit to Dubrovnik was conducted in October 2019, several months after the show concluded its eight-year run, at which time I went on another commercial walking tour, although without interviewing fellow tourists. This means that the research was largely conducted before Northern Ireland's push to firmly associate itself with *Game of Thrones*,[17] and while tourism for the show in general was still figuring itself out.

This resulted in identifying a broad range of tourists with a variety of connections to *Game of Thrones* and the A Song of Ice and Fire book series, as well as capturing tourist activity as it was beginning to coalesce. I interviewed tourists who were specifically at the location or taking the tour because of *Game of Thrones*, in order to focus on the "fan" audience rather than the general tourist audience. Fans with different perspectives on *Game of Thrones* and its locations could be considered: those who had been involved with the book series before the television show debuted or those who had recently begun watching; those who frequently participated in activities with other fans and those who preferred to watch on their own. I do, however, consider all of them fans, following Sandvoss's definition of fandom. There were rough differences between the different groups, with the fans at TitanCon being more likely to be part of "participatory" fandom culture, doing activities such as discussing the show regularly with other fans online and creating fan works, and those interviewed in Dubrovnik being less likely to do such things. However, this does not mean that the fans in Dubrovnik cared less about the program than the ones found at TitanCon, only that they expressed their fandom in different ways.

In all, forty-eight film tourists, aged nineteen to sixty-three, were questioned alone or in small groups. Thirty of these were between the ages of twenty-four and thirty-five, with twelve older and six younger, which is fairly consistent with the known demographics of the *Game of Thrones* viewership. Twenty-three were female and twenty-five were male. Seventeen were from North America, twenty-one were from the United Kingdom, and the remaining tourists were European, which can possibly be explained by TitanCon having a largely British attendance base. If people had the time and willingness to talk in the moment, they were interviewed on site; this practice had the benefit of gauging immediate reactions. Others were questioned nearby, shortly afterward. A minority of volunteers also participated in follow-up interviews via Skype; these tended to offer a more measured reflection on the experience. Two visitors who participated in both the TitanCon tour and a later trip to Dubrovnik with friends from the convention were interviewed after each trip. In all there were thirty-one interviews, lasting from seven to forty minutes. Those questioned have been given pseudonyms to maintain their anonymity.

Participant observation was used to complement the interviews. The Dubrovnik tours were daily walking tours of a few hours in and around the UNESCO-protected "Old Town" with local guides who had also been extras on the show. The commercial Northern Ireland tour was a coach tour along the coastal route outside of Belfast lasting the entire day. The fan-run coach tour, which was considered the second day of an annual convention dedicated to discussing the show and its genre, was also an opportunity to be social with a large group of other fans, thereby allowing me to gain their confidence prior to the interviews. Finally, Fort Lovrijenac was an easily accessible spot for interested tourists who might not have been willing or able to spend the money on the tour or who wanted to experience it on their own time.

Analyzing the data revealed the central role of the imagination in the film tourist experience. More particularly, the interviews showed that there are three main modes of imagination at work: hyperdiegetic, production, and historical. I refer to these as "modes" because, as I will explain, they are different mindsets that tourists can move between, rather than distinct types of tourists. They are presented in this order to first engage with previous research on film tourism, which has often focused on what I have termed the "hyperdiegetic" mode, and then to move from it into the other ways that fans utilize their imagination to experience these places.

The On-Site Imagination

Imagining Westeros: The Hyperdiegetic Mode of Film Tourism

When standing at filming locations of *Game of Thrones*, many tourists imagine the narrative world of the TV series:

> We went swimming down by the city walls the other day, and just thinking, "Oh my God, that's King's Landing! We swam right next to it! That's fantastic!" (Dan, 22, English, in Dubrovnik)

Such a "hyperdiegetic" mode of imagining is, as I've talked about, a cornerstone of many studies on film tourism. Being at a place related to famous story lines encourages a certain narrative imagination. These locations, as Hills discusses, "sustain cult fans' fantasies of 'entering' into the cult text, as well as allowing the 'text' to leak out into spatial and cultural practices."[18] They evoke an embodied sense of the narrative:

Julie is sixty-three and from Toronto, Canada, where she works as a team leader and project manager for a financial planning office. She got into *Game of Thrones* by starting the first book of the A Song of Ice and Fire series, only to have a co-worker start downloading the television show, which she passed on to Julie. The series hooked her in, and she never returned to the books. At the time of the interview she had watched the entire series and was eagerly awaiting the fourth season, which was filming at the time she visited Northern Ireland (which meant she couldn't see some of the places that were normally included on the tour).

She became interested in the series because of her interest in history and historical fiction. She enjoyed reading about World Wars I and II and the works of Ken Follett, citing his medieval historical novel *The Pillars of the Earth* as a particular favorite. The similarities of *Game of Thrones* to "our own history" was what she found particularly compelling about the series, noting the political intricacies, the excitement of living in this more extreme environment, and the effects that the power struggles of Westeros have on its ruling families. The fantastical elements of the series—the White Walkers, the dragons—were something she overcame because she enjoyed the historical-esque political intrigue so much, rather than a draw for her.

Her interest in history brought her to Northern Ireland. Her father had traced their family genealogy back to the country, and she found it important to go and retrace the steps of her ancestors. Because her father was also a geologist, seeing the Giant's Causeway—a unique volcanic rock formation and a UNESCO World Heritage site—was important to her. She was excited to discover a tour that allowed her to see both the Giant's Causeway and *Game of Thrones* locations on the same trip. Visiting *Game of Thrones* locations on a long day of sightseeing kept her interest throughout the day. She also enjoyed that she could take time out of the more obligatory parts of her family heritage trip and see things connected to her favorite TV show—in her words, it made visiting Northern Ireland even more exciting.

It also made her feel more connected to the show. That she had been there, stood on the same ground, in a place with a link to her own family, gave her a more personal connection to it. She had been to other filming locations in a similar opportunistic fashion—she named *The Witches of Eastwick* and *A League of Their Own*, both of which she'd managed to incorporate into family holidays—and found that being there was a way to connect to these favorite narratives and make them part of her own life experience.

While *Game of Thrones* is one of Julie's favorite shows, she didn't participate in the fandom around it—no fan activities other than the tourism—and she didn't read about it online or in print, which she credited to her long hours at work. However, the show is something that she shared with her daughters, who are in their twenties. They all watched the show and discussed it together (and are particular fans of the character Jon Snow). After the interview, she was planning on meeting with them, as they didn't join her on her trip, and showing them all the photographs she took in an organized fashion. Her one complaint was that Northern Ireland didn't have nearly enough *Game of Thrones* souvenirs—there was nothing to bring home for them.

I just take everything. But really, the city walls, the landscapes, trying to picture in my head what King's Landing would have been like. (Josh, 27, American)

Not only can fans imagine that they are swimming in the water outside of King's Landing, as if they were a character on the show; they can also refine their mental image of this location. Physical sensations not transmitted over television can be experienced. The streets of Dubrovnik's Old Town can prompt ideas about what else there is in the city that has yet to be shown on screen.

This imagining of the narrative space of the show is also sometimes performed, either through reciting lines from the show or reenacting its scenes on-site. Visitor performance is, however, constrained and shaped by "communal conventions about 'appropriate' ways of acting as tourists."[19] In Dubrovnik, where the fans mingled with casual tourists in city-center locations, reenactments or recitations were rarer in 2013 and more visible in 2019, suggesting that appropriate ways of acting as tourists in these areas had shifted. Reenactments were expected either where film tourists were isolated or where they formed large enough groups to dominate an area. At the locations visited as part of the TitanCon tour, reenacting scenes was a fundamental part of the experience. This is encouraged by the convention's leaders; previous years also featured such reenactments, and such a repetition can provide the event with a ritual character, as I will discuss in more detail in the next chapter. The organizer of TitanCon is a dedicated *Game of Thrones* fan and felt that reenactments are important to the day and tour experience:

It sort of . . . links people to that location. It's like, "I was there." When you rewatch the scenes, you go "oh yeah, that! I recognize that! I was there!" That's a special feeling for a lot of people. (Thomas, 37, English, TitanCon)

In reenacting, the tourist gets to connect to the location and become part of the show and its world. Reenactment heightens the imaginative experience by incorporating an embodied, physical action, one that suggests a "feeling-as" part of the narrative. It also strengthens the memory of the visit by connecting it to such a special performance.

Hills sees such reenactments and "cult geography" in general as functioning in a near-religious sense, "offer[ing] a physical focus for the cult's sacredness."[20] The religious terminology is also invoked by frequent use of the word "pilgrimage" when discussing fans visiting filming locations, suggesting a certain solemnity toward the location, the travel to it, and the experience there. There are certainly some similarities in form. But, as I discussed in the last chapter, I'm skeptical of this metaphor. One of the reasons for this is that the "sacred" aspect of being at the space seemed to be absent among the *Game of Thrones* tourists. In some cases, it was even directly rejected:

> It's not really a religious experience. It's more . . . it's more a pleasant location and you are there with friends and . . . (Fredrik, 31, Swedish, TitanCon/Dubrovnik individual)

The connection of pilgrimage with sacredness and solemnity overshadows the other ways in which tourists engage with filming locations, even for films and programs of which they are fans.

Rather than a religious moment, the experience for *Game of Thrones* tourists at the sites was usually playful. Reenactments, especially, were described as fun:

> But yeah, we did the reenactment of the Hound, sorry not the Hound, the Mountain, doing the jousting. That was something to do. We've done the crowning of the King in the North, and other bits . . . it just adds to the fun of the day. (Lisa, English, 35, TitanCon)

Tourists involved in the reenactments tended to laugh and joke with other fans while doing so, exaggerating their movements and making humorous references to the narrative. While some might argue this represents a self-conscious distancing from the connotations of fandom as "religious" pursuit, it could equally be said that fans do not need solemnity to feel bonded together. Williams stresses the particular importance of play in fan tourism, which allows for a "looser and imaginative"[21] encounter that opens up the story-world, expanding it further through the creative work of fans in place. A playful approach offers ways of recalling and imagining the world of *Game of Thrones* that allows them to feel part of the narrative and imagine what the world feels like in an embodied manner. That fans know it's a fantasy means that their experience needs to be in some way playful to make sense, play being a way to reconcile the imaginative with the real.

Tourism therefore differs from text-based explorations of a story-world. As Crouch suggests, physical encounters produce a different knowledge, a "multi-sensual" understanding of place,[22] and in this case, of narrative.

Sensual and emotional experience can be as important to the tourist's sense of connection with the narrative as the locations themselves. For example, Reijnders found that connecting with the Dracula story on-site was not only about visiting exact locations.[23] It also meant participating in activities that matched the emotional state of the novel. Experiencing something "like" the story while being in its environment is part of what makes the experience meaningful. Buchmann, Moore, and Fisher, in their investigation of *The Lord of the Rings* tourists, found that connecting to the themes of the text, identified as "fellowship, adventure, and sacrifice,"[24] was as important as being at the actual place. In finding "fellowship" with other tourists at the location, the text "happened," and in experiencing an adventurous landscape, they had a more authentic experience of the text by participating in something emotionally like it.

Game of Thrones fans were no different. Many of the TitanCon tour participants especially enjoyed the concluding activities: group workshops on archery and sword fighting, as well as feasting in a large stone hall in a forested estate. These activities allowed tourists to experience what being in the show's narrative world would be like but in a way that also involved having fun with other fans. They got to play with the general narrative space of *Game of Thrones* alongside others who were equally happy to be part of the moment. As Roesch, Kim, Buchmann and colleagues, and Jones discuss, being with others who "get it" can be an important part of the film tourist experience—perhaps just as much as visiting the locations themselves. The presence of others potentially enhances the experience by reinforcing the importance of the location and encouraging deeper participation in the activities that occur around it. The tourists are not only immersed in the location, but within the community—an important aspect of film tourism that I will explore in the next chapters.

Interestingly, within the TitanCon group, the larger collective of fans present encouraged a more playful attitude toward the text and attendant activities than reported in other studies on film tours. The reenactments were particularly playful. This matches with the somewhat ironic and postmodern tone of *Game of Thrones* toward other medieval-styled fantasy texts (such as *The Lord of the Rings*). As I'll discuss in more detail later in this chapter, *Game of Thrones* presented itself as a more "realistic" take on the genre, without the mythologizing of the past and its society that is often seen to characterize it. Within a group of fans, this presents itself as a less reverent stance in general. Instead of solemnity, they are encouraged to feel how the past "really" was and have fun with the text—which includes jokes, songs, and drinking.

While there are moments of slippage for tourists between our world and Westeros, these were mostly constructed playfully and were somewhat fleeting. Tourists might feel for a moment that they are swimming next to King's Landing, but this is not the total experience with Dubrovnik. Rather than complete immersion in fantasy, the key places instead function as ways to grasp the narrative in greater detail and precision. It is not that Westeros really exists, but rather that it could exist; tourists therefore use these locations to get a physical idea of how it might feel, and play with the thoughts of what it would be like.

Imagining How It's Made: The Production Mode of Film Tourism

Narrative space is not the only thing imagined by *Game of Thrones* tourists. After all, fantasy films and television shows are put together, with considerable effort put into the construction of believable fantastic worlds. This production process is known to be interesting for fans, featuring in the discussions at fan conventions and offered as an enticement to buy DVD sets and other such paratexts. For some, imagining this production process is more prominent:

> Certainly [I enjoyed] looking at the surroundings, and seeing how they turned that into King's Landing. It's not so hard appreciating that when you're in a medieval town like Dubrovnik in the old town, but certainly when you're at this modern hotel that was also just covered in graffiti and windows smashed because it was derelict, and we were walking down and finally found where they'd filmed and I was just thinking, gosh, they brought actors here and everything, and how did they get in, how did they do it, and how did they find this place in the first place? It's quite impressive. (Jane, 41, Scottish, TitanCon/Dubrovnik individual)
>
> Well, you can imagine that you were in the production. [. . .] That would be . . . yeah, it's a sort of dream to be. (Fredrik, 31, Swedish, TitanCon/Dubrovnik individual)

For such tourists, experience at the location is not centered around imagining how the fictional world "is" or how the characters interact with it, but how the show itself is actually made. During their visit they found not only locations that were already shown onscreen, but also ones that were going to be part of the upcoming season. This meant that fans could imagine the work that the production staff had put in to make the scenes they had already appreciated and how the site they were standing on would be transformed in the future. They could, as Fredrik did, imagine themselves as part of the team that created the show they love, a production mode of imagination.

On one level, tourists like Jane and Fredrik exhibit the fascination with the "media world" that Couldry[25] credited with the appeal of the *Coronation Street* set and the Granada studio tour. They are intrigued by the work of the television production team and want to see how they function:

> I've always enjoyed television and been interested in television, and all aspects of it, so I think that's what I'm liking about [*Game of Thrones*], it's just appreciating, from all the people who work on it, from the people who are stitching up the costumes to the lighting guys to everything, it's quite impressive. (Jane, 41, Scottish, TitanCon/Dubrovnik individual)

Jane is an interesting example of this kind of tourism. As a long-term reader of the A Song of Ice and Fire books, she has been interested in the television adaptation from when it was first announced, even watching the filming of the pilot, which was nearby her home in Scotland. She already had knowledge of the show's story lines (at the time of the interview) and a sense of the narrative locations involved. This gave her a different perspective on the show compared with other television programs that she had been interested in, and she found herself wanting to know how the show was made—the work and effort that went into bringing the Westeros she had always imagined into real life. While she had always been interested in television production, with other series she did not want to lose the tension of not knowing what would come next by finding out the behind-the-scenes stories. With this concern lessened for *Game of Thrones*, she could focus more on production practices.

For Jane, visiting *Game of Thrones* filming locations, as well as attending fan conventions like TitanCon, is a chance to see the way in which the "special," usually concealed world of the media operates. In doing so, Jane experiences hints of the "deep backstage" of actual film production[26] that is rare and therefore exciting. This gives her a deeper understanding of the actual world that surrounds the show's making, rather than its fantasy world, one that helps her to appreciate the show in a new way:

> How do I feel about it? Satisfying, I suppose, to sort of figure out that gosh, look at the lengths they've gone to, etc. Another situation was a surprise location that we found when we were on the island of Lokrum, one of our group asked the barman, has there been any filming around

here for *Game of Thrones*? And he said, oh they used our storeroom. And we got taken into the storeroom, very kindly, and we were clambering over bits of polystyrene and plastic on the floor and there's broken umbrellas and chairs and things, and we suddenly realized we were in this beautifully painted hall which was used in one of the scenes. And we could only see that when we set the flash off on our cameras, because it was so dimly lit, and it was amazing, the artwork and things on the wall. And how the *Game of Thrones* production team discovered about that place, I have no idea. (Jane, 41, Scottish, TitanCon/ Dubrovnik individual)

Couldry sees the audience's interest in television production as a way of reifying the "specialness" of the media world, one that is potentially more interesting, creative, and fulfilling than their own. Yet, locating the "media world" as a single entity obscures something significant: fans are interested in the making of *Game of Thrones*, not necessarily a more generalized television production process:

Well . . . to get inside information, to maybe . . . I have an eye for detail so I could maybe spot some errors before they get to the screen. (Fredrik, 31, Swedish, TitanCon/Dubrovnik individual)

Fredrik, a friend of Jane's, does not want to work on just any television show; he wants to work on *Game of Thrones*. Like Jane, he is a long-term reader of the book series, as well as a fan of the television show, and his interest in working on the production is because of this fandom. Rather than an interest in television production generally, he wants to either learn things about the program before it airs or suggest ways of "fixing" poor episodes. By imagining himself as part of the production staff, he explores the idea of assisting the show he wants to be part of. While it might also be exciting to be part of the "media world," it is most exciting because that media world makes *Game of Thrones*. This is consistent with Kim's study of tourists at the filming location of the popular Korean television drama *Daejanggeum*, where involvement with the narrative was credited to higher interest in "behind-the-scenes" experiences there.[27] Media production in itself might be interesting, but emotional involvement in the narrative makes it more meaningful and imaginative.

The production mode was also a major part of the Dubrovnik commercial tours, making up many of the anecdotes shared by the guides in 2013, when the tours were relatively new, and in 2019, when *Game of Thrones* tours were the most popular walking tours in the city. The tour guides, who worked as extras in the off-season, provided a great deal of "behind-the-scenes" gossip to fans, who seemed eager to hear about how the production of the show actually functioned in the small city. The presentation of this information was designed to give the tour participants a sense of "being there" during the creation of the show—of knowing how the cast and crew moved through the space, what the cast was like in their downtime, or even what it felt like as a Dubrovnik resident, seeing the filming take place around you and feeling part of it. The link to Dubrovnik itself was also crucial here, particularly in comparison to Jones's exploration of a similar *Walking Dead* tour[28] where extras served as the tour guides, but the story of the location itself and the experience there is less prominent (and in some cases, restricted by AMC policy).

In both the 2013 and 2019 walking tours, there was stress on the way in which the production team and actors adapted to and spent time in Dubrovnik itself, and how the "locals" found themselves involved in production at different levels. The accessibility and friendliness that the extras/guides claimed the production had during filming, which took place in the touristic off-season, is starkly different to the divide between extras and "core" cast that Jones reported on, showing the different sort of approaches that tour guides can take in setting up the production mode of imagination and connecting it to local concerns. The sort of gossip used in Dubrovnik not only encourages tourists to think of Dubrovnik and its people as special, perhaps as a response to concerns about *Game of Thrones* taking over Dubrovnik's identity,[29] but fosters a sense of imaginative inclusion in the production, and is promoted as a reason for taking a walking tour rather than seeking out locations for oneself. In participating, the fans can (ideally) imagine themselves as part of the group that puts the show together, included in the creation of something they find special.

However, tourists can also explore for themselves. Jane and Fredrik were among a group of friends who met at the TitanCon convention and subsequently took a trip to Dubrovnik together to see filming locations there. Since *Game of Thrones* was an ongoing series at the time, over the course of their trip to Dubrovnik they also found remnants of filming locations for episodes that had yet to air. They are therefore assisted in imagining what would happen in future episodes. They had some idea of what would take place and were excited to see how key ideas would be transferred to the screen. The nature of the ongoing adaptation—both knowing and not knowing what would come next in the series—gave the production elements the sense not just of relics, but of the future. It also gave the group a sense of exclusivity, compared with other fans: they had seen something special and new about the production.

Additionally, fantasy fictions like *Game of Thrones* are sometimes "considered [as] not 'serious' or 'mature'"[30]—the preserve of children, teenagers, or adults who have not grown past what they enjoyed at early stages of their life. Fantasy fans are often equated with "geek" or "nerd" stereotypes,[31] an unappealing image. Some *Game of Thrones* fans are therefore keen to stress that their show is of high quality:

> [*Game of Thrones* is] just a really well-written character drama that happens to be swords and dragons and . . . as I said again, dead geeky, but similar to how I feel about Battlestar [Galactica]. (Wayne, 32, English, TitanCon)
>
> Every once in a while you come to a show that really takes the time to create engagement and a plot line that really fits in with the story and it, you feel more connected with the characters. (Afra, 28, American, Dubrovnik tour)

Calling attention to high production values, represented by the creative work put into finding and transforming locations, is a way to stress quality. The production is not something that was cheaply put together or lazily executed. Rather, it is something that HBO (itself a self-proclaimed arbiter of a certain taste level) deemed worthy of a large budget to hire high-quality set designers, location scouts, and state-of-the-art computer-generated imagery. In admiring and envisioning this effort and skill, film tourists show that the program has rewarded the time and effort they have put into being its fans. By visiting its locations, they can better imagine the skill involved in its production, and how that might maintain its future appeal.

Imagining What Happened: The Historical Mode of Film Tourism

A third mode of imagination pursued by film tourists visiting sites connected to *Game of Thrones* can be described as historical. For certain visitors, experience of a filming location is only partially, and in some cases not at all, connected to the narrative or production world of *Game of Thrones*:

> I think in terms of, uh, some of the older buildings and so on, it's more just a sense of wonder as how these things could be constructed and so

on. I didn't really associate it as much with the show, more just with the
buildings themselves. It was kind of . . . I was almost divorced from it.
(Duncan, 28, Scottish, TitanCon)

For these tourists, *Game of Thrones* is only a portion of the imaginative experience of the locations they visit. Although it brings them to the filming
location, while there, and while thinking about it afterward, they imagine
other aspects, in particular the historical narratives on which the specific
place identities are built. In tourism research, film is usually considered
more of an incidental driver for tourists rather than a direct one, with filmic
destinations primarily functioning as an image enhancer, or even changer,
for regions rather than a primary motivation to visit.[32] Those who actually
visit the filming locations are still seen as a more niche audience within the
general tourist market, as most tourists might become interested in visiting
a region because of its depiction on screen and its resulting emotional resonance but will not necessarily seek out the exact spot of filming.

The tourists interviewed here, however, while having a range of reasons
for visiting Dubrovnik or Northern Ireland itself, all made a point to visit the
specific locations used in *Game of Thrones*, making them part of this niche.
Intriguingly, we can see in many visitors to filming locations a companion to
the situation described by Croy[33] and others, in that *Game of Thrones* creates
a frame to begin experiencing the place, in terms of both the exact location
of filming and the general region around it. Once at the location, their sense
of it as being part of the show is less important than a feeling that it is a
space worthy of investigation on its own terms. Facilitating this approach,
in both Dubrovnik and Northern Ireland, many *Game of Thrones* locations
are altered through decoration and computer-generated imagery. The real
places contain elements that are on the show, but not everything that
appears on screen. This makes a direct comparison impossible and forces
the tourists to understand differently:

It's just . . . because most of these places, they don't really look like how
they look on the screen anyway, they're not really that recognizable,
so you just sort of enjoy them through what they are, for the spectacular view, rather than for the fact that they're in a show. (Melanie, 31,
English, TitanCon)

The locations are not what Melanie (who, like Duncan, came to Northern
Ireland for the *Game of Thrones*–focused TitanCon experience) imagined

them to be, so, alongside fellow tourists, she has to make other meanings to enjoy the day: either consider the fine production work used in spotting and creating the locations, or appreciate these sites as part of the beautiful, previously unknown Irish landscape.

Of course, however, some locations do resemble their on-screen appearance. The arching beech trees in Northern Ireland known locally as the Dark Hedges were, in *Game of Thrones*, a striking element of the "Kingsroad" that connected two regions of the Seven Kingdoms, traveled by one of the leading characters at an integral point in her story line. Such iconic locations make up an important part of the commercial *Game of Thrones* tour in Northern Ireland, and similar locations do the same in Dubrovnik. The trees, however, have existed longer than the show, and they come with their own narratives:

> The only thing I've done since I've got home is look up the Dark Hedges. [. . .] That hedge was built as a spectacular entrance to their home, and probably in 1750 or something? I mean, I'm just fascinated by those hedges! (Julie, 63, Canadian, Northern Ireland commercial tour)

Game of Thrones sparked Julie's initial interest in the hedges, but for her, it was used as an impetus to learn the "real story" and imagine the lives of the historical Stuart family. Her interest in history is what had brought her to Northern Ireland in the first place—she went in order to encounter her family heritage, which her father had traced back to Northern Ireland. That *Game of Thrones*, one of her favorite TV shows, was filmed there was a happy coincidence, and she made a point of taking the daylong tour of *Game of Thrones* locations, which provided some emotional relief from the more serious obligations of her trip. Yet it was still history that ended up occupying her thoughts after she went home. Filmed in locations that are embedded in the historical structures and landscapes of their respective countries, the show therefore acts as an entry point to the exercise of historical imagination. The possibility that such places might be interesting beyond their connection to the series means that many tourists pursue their visits as a way to understand not only *Game of Thrones*, but "real" places as well. Tourists can use what they already know about the show as a starting point to find out what they do not yet know about the existing place narratives of Northern Ireland and Dubrovnik.

These existing place narratives were well set up to do what Lundberg, Ziakas, and Morgan refer to as "align[ing] on-screen narratives with destination qualities,"[34] which they identify as a key strategy for places that want to successfully market their film-related destinations. Julie and many of the other interviewed tourists were particularly interested in their show's connection to history and its sense of "historical realism." Unlike Jane and Fredrik, who had a longer-term interest in fantasy fiction, Julie's primary interest in the series was in the way it depicts history. She describes the appeal of *Game of Thrones* for her:

> It's mainly history. That's what I liked about . . . you know, it took me a while to get into the book. When you have White Walkers and, it's kind of the equivalent of our zombies, you know, or maybe that's the appeal for the younger generation? That would not be my main draw for that series. I think it's more the struggle, I mean it's the story of, it's our history. It could be anywhere, struggling for power, right? [. . .] Is this the way these people lived? You never knew if you were going to live or die or live to see the next day, and . . . you just can't imagine, what it was like. (Julie, 63, Canadian, Northern Ireland commercial tour)

While a sense of the medieval is common in high fantasy, part of *Game of Thrones*'s claim to distinction as a "quality" program is that it has successfully created a sense of historical authenticity. Despite its fantastical elements, the show prompts audiences to imagine what the historical past was "really" like and presents itself as a fictionalized but not inauthentic version. For many tourists, learning about the "real" history of the place in which it is filmed affirms the imagining of the past through the show as valid. Tourists across age groups and fan types expressed an interest in history, making it one of the most common linkages between fans, although older fans were more likely to pursue this interest outside of *Game of Thrones*. Tour guides knew this and pleased their audiences by discussing the real history of key locations. Historical structures both shown and not shown in the show were explained, as tour guides discussed how they were built and used. In describing Dubrovnik's past as the Roman Catholic Republic of Ragusa, one guide also pointed out where the characters of *Game of Thrones* would have lived and worked, if they had lived there, while another described Ragusa's role in European history by comparing it to the more subtle, scheming characters of *Game of Thrones*. Through such framings, tourists can imagine what historical Dubrovnik (and potentially King's Landing) was like, a strategy well in line with the Croatian tourism board's focus on links to Western Europe rather than the more recent war.[35]

The appearance of historical imagination among film tourists is not uncommon, as suggested by Lundberg and colleagues, who consider it the first of their six strategies for promoting film tourism destinations. New Zealand, one of the most well-known locations of film tourism, deliberately drew on its existing branding as a place of "raw nature" to enhance its suitability to not only be the filming location, but to be a tourist destination, for the epic fantasy landscape of *Lord of the Rings*'s Middle-earth.[36] Buchmann and colleagues show that many fans on the *Lord of the Rings* tours wanted to "see New Zealand" in addition to the filming sites, but "most if not all participants continued to interpret New Zealand as a green and friendly place without any significant problems."[37] They viewed the country and their experience of touring in a manner that aligned with the myths of the "timeless" rural nature of Middle-earth, despite also visiting urban environments. Similarly, the Harry Potter tourists studied by Lee interpreted Britain by conceiving a magical world alongside, but hidden from, everyday reality. Their tour stopped not only at Harry Potter locations but also at ones associated with local legends and mystic sites to create a pseudo-historical, "magical" picture of the United Kingdom.[38] It is also not necessarily tied into the narratives promoted by tourist boards. Focusing on science fiction fans in Vancouver, Hills[39] and Brooker[40] found that film tourist versions of the city go beyond the "regular" tourist spots, in search of any location that "replays the 'hiddenness' of *The X-Files*' own tropes and secrets."[41] By focusing on the "secrets" behind the every-city façade of Vancouver, it becomes a mysterious place waiting to be revealed by the traveler. Garner situates this way of interpreting place narratives as part of the "intertextual spread"[42] of transmedia properties, which intersect with other narratives as they expand into different forms of media. He argues that this is particularly the case with tourism, as the iconography of the transmedia world connects to related narratives in our reality, transferring some of its meaning and affective power to these spaces.

The fans here did, however, want the "real" history, not only the textual one. Just *Game of Thrones* would have been less satisfying than *Game of Thrones* and the actual stories behind the locations. A complete overwriting of Northern Ireland with Westeros and Dubrovnik with King's Landing is not necessary, and for some tourists (and residents), it would have been detrimental to their experience with these places. That they are two things is important to the imaginative experience fans have and want in regard to *Game of Thrones*. Being there means getting exposure to both historical and fictional narratives, which for many was a reason to do the tours in the first place. The show is a frame, not a covering.

Still, historical imagination is never neutral. It takes a particular form depending on the text being de-mediated. This separates film tourists from "regular" tourists, in that even though they are interested in aspects of the location that are not strictly part of filming process, the contours of their imaginative experience are shaped by a notion of history provided by popular culture. Just as *Lord of the Rings* fans see New Zealand as timelessly pastoral and spectacular, so *Game of Thrones* fans frame Dubrovnik and Northern Ireland as part of a mythic-medieval world. In Dubrovnik, this sense is reinforced by the city's well-preserved historic status:

> It's everything we thought it would be. [. . .] It's as beautiful as people said it was. The history, the culture, the preservation of everything. It's all there. (Josh, 27, American, Dubrovnik individual)

In this way, visiting filming locations confirmed the association of Dubrovnik with a sense of beauty and the long-ago past, something generally encouraged by existing fuzzy historical narratives around Dubrovnik that rarely dwell on recent times.[43]

In Northern Ireland, however, impressions of the country were often challenged through the visit:

> I was a little nervous at first, of coming, wasn't I? [. . .] I imagined it to be more of towns [. . .] and this is going to sound bad, but less . . . taken care of? But it's not. It's actually a lot of greenery and it's very beautiful. (Liz, 34, English, TitanCon)

Many tourists generally associated Northern Ireland with the Troubles: the sectarian violence that gripped the country in the second half of the twentieth century, which has been the focus of most of its media representation.[44] *Game of Thrones* locations offer an alternative, more positive image of Northern Ireland, as a green and aesthetically pleasing land—one filled with forests, castles, and stunning wild views. It is, as Çelik Rappas and Baschiera discuss,[45] a way to connect Northern Ireland to a different sort of heritage narrative, one that is both cooler because of its association with *Game of Thrones* and more appealing because of its connection with a more romantic, aesthetically pleasing history.

The historical mode of film tourism is therefore not a new finding per se, but one that should be thought about in the context of fandom. It is not just that *Game of Thrones* promotes a particular narrative of Dubrovnik and

Northern Ireland, but that this narrative should be seen in terms of these places' existing fandom. The use of *Game of Thrones* as a frame or jumping-off point facilitates a specific imaginative experience that connects to what fans like about the show. It does not entirely take over the actual place identity, but it does provide a new path for tourists to go down when exploring.

Conclusion

Using the idea of the imaginative experience opens up our understanding of how fans/tourists can pursue multiple readings of a site and its relation to its associated text. In examining a series of interviews with *Game of Thrones* fans in Dubrovnik and Northern Ireland, I aimed at showing how this imaginative experience takes place in practice and which modes of it can be distinguished. As the interviews showed, fans of *Game of Thrones* involve their imagination in the on-site experience in three distinct ways.

In the hyperdiegetic mode of imagination, fans primarily draw on the show's narrative, imagining places in which the show happens and envisioning themselves as part of its story. Being there not only offers these fans the possibility to explore the space where it "all took place," but also allows them to walk outside the set and discover the surrounding "hyperdiegetic space" of *Game of Thrones* that has not (yet) been used as backdrop for the filming. Wandering through the streets of the old city center of Dubrovnik, for example, the *Game of Thrones* fan is able to imagine that they might run into Cersei at any moment, and picture their own role in the sprawling narrative.

Film tourists also use their imagination to reconstruct the filming process. Being there helps fans form an image of the technical challenges as well as the creative performances of the production team. In addition to admiring the work that was done to the location, some film tourists also liked to envision themselves as part of the team that put it all together, something encouraged by the tour guides. Whereas the hyperdiegetic imagination focused on feeling like one is in the world of *Game of Thrones*, the production imagination was aimed more at retracing the steps of the production team, the actors, and anyone else who was involved in putting *Game of Thrones* together in our world.

Additionally, the *Game of Thrones* narrative can be used as a frame to understand the "real" history of the places that fans visit. The association with *Game of Thrones* can be a jumping-off point to explore the rich history of the respective places. That said, the frame of *Game of Thrones* left its stamp on the kind of histories people were interested in. For example, in line with the pseudo-medieval profile of *Game of Thrones*, film tourists in Northern Ireland expressed an interest in the long-ago history of this region. The more notorious association with the Troubles stayed in the shadows, a side effect likely welcomed by the local tourist boards, in their search for a more positive image of this part of the United Kingdom.

Fans are fluid in their imaginations as tourists, and therefore I do not intend this to function as a typology of film tourists. While any one individual might favor one mode over the other, as I show, on location the modes tend to blend into and influence each other. Indeed, the same tourist might experience them all at different times or even simultaneously; the modes are not so much distinct types of tourists as ways of imagining that any fan might slip into and out of during a visit. I also found each mode occurring among the different groups of tourists, with some of the highly organized and traditionally active TitanCon fans interested in the history of Northern Ireland, and fans walking in Dubrovnik, who had never interacted with other fans, dreaming of what it would be like to live in King's Landing. There is, moreover, no clear formula that can predict when a certain mode will dominate. We might expect places that strongly resemble those from the show to inspire hyperdiegetic imaginings, but this is not necessarily the case.

Each mode shapes the others. Fans who have been to Dubrovnik in the summer may think of King's Landing as hotter than they did before. The tour guide's description of the *Game of Thrones* actors and crew hanging out in Dubrovnik, impressed by its preserved architecture and friendly locals, suggests that it is indeed a special place. *Game of Thrones* makes its filming locations seem more vibrant and exciting, but those places also lend the show a renewed sense of meaningfulness because tourists become aware of their physicality and "real" historicity. In a similar vein, while the "real worlds" of Northern Ireland and Dubrovnik are not completely obscured by the notion of Westeros, they are not left unaltered. Rather, fans start to experience them more like their televised counterpart, and they acquire a mythic-medieval glamour. Even if a tourist is more interested in one or the other, they all work together to create the imaginative experience of the location.

Tourists also adapted these modes to fit with *Game of Thrones* specifically, in that the way in which the modes were experienced is influenced by *Game of Thrones* as a text and the way the tourists feel about it. Its "realistic" take on the medieval fantasy genre shaped the way in which tourists interacted with the actual history, while also encouraging a less reverent stance toward hyperdiegetic imaginings. This is consistent with other studies on film tourism, such as those that describe *Lord of the Rings* fans finding New Zealand to be a land of epic adventure and fellowship[46] and *X-Files* fans in Vancouver[47] delighting in "discovering" the more hidden parts of the city. It suggests that, while the different modes are likely applicable across different examples, the way they are experienced by tourists will be shaped by the text.

While here I focused on the immediate (imaginative) experience of film tourism, there is also some indication of what role these places play in *Game of Thrones* fandom. Because the series was fairly new and ongoing at the time of research, there was much that the fans, even those who had read the book series it is based on, did not know about it. Visiting these places is used to explore the show and its potential. Tourism fills in the developing picture of both the story-world of *Game of Thrones* and its creation. Compared with the cases I analyze in chapters 3, 4, and 5, which are more established fandoms for completed series, *Game of Thrones* tourism at this stage was very much about this kind of exploration and knowledge gathering. It is a unique and enjoyable way for fans to learn more about a show they are coming to love, and to bring this knowledge back home to friends and family (providing a bit of bragging rights along the way). The role of these places for fans is therefore mostly as a site of learning and play.

What I primarily argue in this chapter is that the study of the experience of film tourism needs to take into account not only the affective nature of the text being visited, but also the different ways this affective relationship can play out in the tourist imagination. Imaginative experience is not uniform, even among tourists focusing on the same text. However, this was just one visit—a place for fans to go, explore, and come back from, perhaps never to encounter again. What happens when the place and the fandom are more intertwined? In the next chapter, we look at what can happen when the connection runs deeper.

The Prisoner

Fan Homecoming and Placemaking in Fandom

Most studies of film tourism focus on the same sort of case: a then-popular film or television show, at the moment of its popularity, focusing on the "here and now" of film tourism as it happens, fresh in the mind of everyone involved. I'm no exception here—the other case studies in this book focus on a then-ongoing TV show and three new attractions, after all. In general, in this field, less attention is paid toward filming locations that have endured over time, with notable exceptions coming from studies on *The Sound of Music* and its locations in Austria,[1] tourism connected to the James Bond series,[2] and the *Blade Runner* locations in Los Angeles.[3] However, even in these cases, the focus is on one-off touristic encounters with these places; while they might have endured in the public consciousness, there is little consideration of how the relationships that individual tourists have with these places have developed over time.

This continues as we move out of strictly film or television tourism into media-related tourism more generally, even for those media objects that have considerable fan followings. Many locations, such as filming locations, authors' houses, recording studios, and the like have been theorized in terms of "fan pilgrimage," in that fans visit these "sacred" locations associated with their objects of fandom much in the way that more traditional pilgrims visit locations affiliated with religion. The term is popular among both scholars and fan tourists themselves, and visits to fan-related places are frequently described in this fashion. As I've said before, I don't generally like using the term "pilgrimage" for film tourism. Despite this, it is no doubt a useful metaphor for some film tourism, and one that has considerable currency in many circles. It also exemplifies the way in which film tourism is usually discussed: brief, albeit powerful, moments of interaction between fans and places. The fan-pilgrim goes to the location, experiences the desired connection, and returns home, enriched by the experience but without a need or desire to return. This portrayal of film tourism as both an individual and temporary state overlooks the longer-term experiences that fans, and fan groups, can have with these important places, much as pilgrims who make their trip again are often overlooked in the popular conception of pilgrimages.

One example of a longer-term relationship is that of the fandom around the television show *The Prisoner*. For fifty years, fans of the program have been visiting its primary filming location of Portmeirion, a holiday village in North Wales. Their visits support a *Prisoner*-themed shop and a fan-run yearly convention. Some fans visit Portmeirion regularly, incorporating it into both their fan practices and their lives. It is more than just somewhere to see once; it is somewhere to go often, a place that becomes familiar. Being able to not only visit Portmeirion, but continually return, has had an influence not only on individual fans but also on the forms and practices of *Prisoner* fandom as a whole.

It is that influence that I investigate here. In this chapter, I ask what the potential long-term role and significance of place is in fandom, through a case study of *The Prisoner* and Portmeirion. This builds upon the previous chapter by looking at the long-term effects of film tourism on fan practices, and how these places are experienced in multiple visits over a longer period of time. While *Game of Thrones* fans were just getting to know it and its world, *Prisoner* fans have been living with them for decades. Here, I investigate the way in which the relationship between *Prisoner* fandom and Portmeirion has been built over the years and what Portmeirion's role in the fandom is today, expanding the ways in which the role of film-related places in fandom are understood.

Place and Fandom

As I talked about in chapter 1, objects of fandom matter to fans—as Sandvoss suggests, "the object of fandom, whether it be a sports team, a television programme, a film or pop star, is intrinsically interwoven with our sense of self, with who we are, would like to be, and think we are."[4] In essence, this means that we identify who we are through our objects of fandom—they possess traits we consider important, and in being a fan of them, we identify ourselves with those traits. Williams builds on this, discussing how being a fan can give a sense of "ontological security" by providing a stable sense of self-identity not only through the object of fandom itself, but through the community of other fans as well.[5] The object and the community make up a part of the fan's life and provide tools to create a self-narrative and a stable sense of self. While this, like other examples of identity formation, is often talked about as part of childhood, adolescence, and/or early adulthood, as Harrington and Bielby[6] and Harrington, Bielby, and Bardo[7] explain, it can also be an important resource to draw on as fans age. Fandom can provide a continuing structure throughout the fan's life, providing a framework to navigate different life stages while maintaining a stable sense of self.

Visiting locations associated with the object of fandom can therefore be a powerful emotional experience. When a place and an object of fandom have a solid, and especially a long-term, affiliation, the place provides the fandom with a "stable, highly visible, physical anchor in the real world."[8] It is somewhere that a fan can visit to honor their fandom clearly and specifically. Structures and locations provide a tangible, enduring stability to something that would otherwise be potentially disposable.

Despite this, most investigations of even long-term fan-affiliated places look at these visits as single trips (this book included). Yet as fandom is a lifelong pursuit, it is likely that many fans visit these locations more often. This creates a different sort of relationship between the fans and the place than that of a single visit. As Tuan discusses, a "sense" of a place is made up of repetition—of revisiting enough to know and understand it, and to imbue it with meaning built over time. Tuan discusses this in terms of the familiar paths taken in the home, stating that "as a result of habitual use the path itself acquires a density of meaning and a stability that are characteristic traits of place."[9] This idea is further elaborated by Seamon in his concept of the "place ballet," the interaction of multiple people's habitual, frequently noncognitive movements in time and space, which creates a "climate of familiarity which grows and to which they become attached."[10]

For Tuan and Seamon, the sense of a place, of understanding it as a distinct area with meaning, is based on repetition, in doing things over and over again in the space and becoming familiar with it and the others who do the same. While Tuan does note that some seemingly fleeting experiences with place can make an impression on a person's life as large as or larger than that of a long-term residency,[11] deep knowledge of a place requires repetition and reoccurrence. The importance of such repetition and its resulting familiarity returns in most discussions of place attachment.[12] Therefore, while a single pilgrimage might be powerful, a repeated engagement with a media-related location would give the fan a different emotional experience. Fans of *The Prisoner* who are regular and long-term visitors to Portmeirion are therefore likely to have a different sense of it than *Game of Thrones* fans who make a single visit to Dubrovnik have of that place.

For this reason, I propose the concept of the "fan homecoming" to describe such visits, as an alternative to the more frequently studied "pilgrimage." A "fan homecoming" is here defined as a return visit to a familiar fandom-related place. Compared with a single visit, a homecoming involves visiting a place that the fan has come to know, where they have made pathways and place ballets that give it a climate of familiarity and comfort. Through continued visitation, a special location is made into a true place—an area with deeply felt meaning, one that the fan can feel truly attached to. As Sandvoss suggests, the object of fandom itself already functions as a sort of *Heimat* or emotional home for many fans,[13] and via the fan homecoming, this emotional home gains a physical counterpart.

Like the "second home" of a rural cabin or beach house,[14] this fandom home operates as a familiar retreat from "everyday life," a way to "disengage" from the mainstream and connect to a more genuine self while still being in a comfortable and familiar environment. Much of the discourse around second homes is built around a dichotomy of urban and rural, with the urban representing the stressful and inauthentic "everyday life" and the rural as the more emotionally authentic "escape." We can map this onto the concept of fandom as an "alternative social community,"[15] built around interests, attachments, and enthusiasms that those in the fan's "mainstream" life don't share or understand. The fan homecoming gives this alternative community a specific place, one that they can travel to bodily in order to feel both included and free. Given a regular return, the place potentially takes on significant memories for this community, becoming an integral part of its sense of cohesion.

This experience, I believe, is at the heart of the relationship that fans of *The Prisoner* have with Portmeirion. This is not to say that every *Prisoner* fan who visits Portmeirion does so as a homecoming, but that the potential for such a relationship is there, and in many cases, actually is.

Following an overview of *The Prisoner* itself, Portmeirion, and the study's methodology, I show how this works in practice, illustrating how fans of *The Prisoner* become attached to Portmeirion, and how its role as "home" of the fandom is significant for the fans and the fan community that created it. In doing so, I further develop the ways in which the relationship between fandom and place can be conceptualized.

The Prisoner is considered one of the quintessential "cult" television shows, maintaining a dedicated fan base and critical acclaim in the more than fifty years since its original airing. Originally airing on the British commercial channel ITV in 1967, the show centers around the titular character, played by series creator Patrick McGoohan, a spy (we think, although this is never made clear) who resigns from his post and returns to his London home only to be drugged, kidnapped, and taken to a mysterious place known as the Village. There, he is assigned a new name—Number 6—and interrogated by a rotating cast of Number 2s as to the cause of his resignation, in increasingly bizarre and surreal ways, while trying (and failing) to escape his confinement. As discussed by Short[16] and Hanna,[17] the show's high production values, allegorical themes, and surreal narrative and structure helped to define cult television, enshrining *The Prisoner* as an important part of the medium's history and creating the sort of modest in size but active and vocal fan culture that would define television fandom. Well-received rescreenings in the 1970s, 1980s, and 1990s, in both the United Kingdom and the United States, as well as elsewhere in Europe, created fresh interest in the program and cemented its place as an important countercultural touchstone.[18] An appreciation society, Six of One, was founded in 1977 and continues to this day, producing fan-written magazines and organizing a yearly convention called PortmeiriCon. (While Six of One spells the convention as Portmeiricon, I refer to it here as PortmeiriCon to better visually differentiate it from Portmeirion itself.)

Textually, its "endlessly deferred narrative,"[19] a feature of cult television texts that is said to encourage the more fannish practices of its audience by opening it up to multiple interpretations, is built on a great number of questions, from Number 6's original name and position to the ownership and purpose of the Village, none of which are answered by the conclusion of its seventeen-episode run. Indeed, the only question definitively answered by *The Prisoner*'s final episode was that of the Village's filming location— Portmeirion, on the coastline of North Wales. A holiday village and hotel designed by Anglo-Welsh architect Sir Clough Williams-Ellis, first established in 1925, Portmeirion had been used as a filming location for McGoohan's prior series, the more straightforward spy program *Danger Man*, where it had stood in for Italy and other "exotic" European destinations. According to fan lore, this visit inspired McGoohan to set the vague idea he'd had for a show about "a man in isolation" there, shaping the tone of the eventual program and the future of Portmeirion itself.

Described by Williams-Ellis as a "home for fallen buildings,"[20] Portmeirion's jumble of architectural elements, Italianate design, colorful paint scheme, and bucolic coastal setting gave *The Prisoner* an instantly iconic look, especially when it was rebroadcast in color. The last episode of the series revealed Portmeirion as the Village's primary filming location, and the connection has endured to this day, with the two frequently connected in popular culture and references to the show appearing in much of the media coverage of Portmeirion.[21] It features a *Prisoner*-themed shop in the building that had been Number 6's house, in addition to a Number 6 café, and it hosts not only PortmeiriCon but also Festival Number 6, a popular music and arts festival that makes use of *Prisoner* iconography. While Portmeirion's draw as a tourist destination is not entirely built on its connection to *The Prisoner*, for those familiar with the program the association is strong, and one that is regularly reinforced by popular culture and, recently, Portmeirion itself.

Methods

This study is primarily based on sixteen semi-structured interviews with fans of *The Prisoner* who have visited Portmeirion multiple times. While undoubtedly many fans of *The Prisoner* who visit Portmeirion do so as a one-off visit, I focus here on the more long-term visitors as a contrast to prior film tourism research, including what I show in chapters 2 and 4. The majority of these interviewees are either current or former members of Six of One, with one interviewee having never participated in the group. Interviewees were recruited from the Facebook group "The Prisoner and Portmeirion," the *Prisoner* fan site "The Unmutual," and the 2016 edition of PortmeiriCon, resulting in a mix of volunteered and solicited interviews with long-term fans who have a range of relationships to Portmeirion and the broader fandom. Six women and ten men were interviewed. The youngest interviewee was thirty-three years of age and the oldest seventy-one, with the majority in their forties and fifties. Twelve of the interviewees were English, three were from the United States, and one was from Northern Ireland, which is relatively consistent with the demographics of both Portmeirion and PortmeiriCon, which market themselves primarily to British holidaymakers and fans of this particularly British show, although some visitors or attendees were from elsewhere in Europe.

Interviews lasted from thirty to eighty minutes and were conducted either on-site in Portmeirion, via telephone, via Skype, or in one case as text messages over Facebook Messenger per the request of the interviewee. This range of settings was based on the availability and desires of the interviewees, with the goal of making the process as comfortable for the interviewees as possible and to allow sufficient time to fully explore the topic. The interviews proceeded in roughly the same fashion, and their transcripts were treated the same, although the on-site interviews tended to be shorter than the Skype or phone interviews. Interviewees were asked about their fandom of *The Prisoner* and how it has developed; their experiences, past and present, in Portmeirion; their feelings about Portmeirion; and the role they feel Portmeirion plays in both their fandom and *Prisoner* fandom as a whole. Interviews were transcribed and read several times before coding in Atlas.ti, and were then reread in light of this coding to develop the thematic analysis. All names have been changed to respect the respondents' privacy.

The interviews were supplemented by three other sources of data. The first was participatory observation at Portmeirion over two visits, including one during the 2016 PortmeiriCon. During this visit I participated in PortmeiriCon activities, including attending discussion sessions, a tour of Portmeirion, and social events and participating in the larger-scale reenactments on-site. I disclosed my status as a researcher and received permission to attend in that role from the convention organizers, in order to approach this private fan event respectfully. Second, I analyzed several bodies of fan-produced and targeted writing thematically via a close reading, including both contemporary and archival material. These writings consisted of recent editions of the Six of One magazine *Six 4 Two*, four issues of the fanzine *The Green Dome*, published in 1980, issues 1 to 22 of the fanzine *Free for All*, published by first the Liverpool and then the Shrewsbury chapter of Six of One in the 1990s, and the regularly updated fan website The Unmutual (theunmutual.co.uk). Third, I reviewed fan discussion on the public Facebook groups "The Prisoner" and "The Prisoner and Portmeirion" and the alt.tv.the-prisoner newsgroup (active in the 1990s–2000s and accessed through Google Groups), with approximately 100 posts in both Facebook groups together and 116 in the newsgroup relating specifically to Portmeirion pulled for further close reading. These data sources were used to supplement and contextualize the data gathered from the interviews, providing important background information as well as confirming certain points brought out in their analysis. Through these methods, I developed the following picture of the role that Portmeirion plays for the fandom of *The Prisoner*, and the way in which these long-term visitors experience the place today.

Michael is fifty-two and works as the director of press for an English university. Originally from Northern Ireland, he got into *The Prisoner* via his friends in the science fiction community there in his early twenties.

Belfast in the 1970s and 1980s was a heavily divided place, and in the science fiction community Michael found a refuge from the conflicts and a group of like-minded people, many of whom remain friends to this day. At the time, he considered himself more of a literary science fiction fan, rather than a television or "media" fan, but he trusted his friends' recommendation. *The Prisoner*, with its allegorical themes, anti-authority messages, and general uniqueness, drew him in, and he became a dedicated fan of the series as it re-aired in the early 1980s. As someone already involved with organized science fiction fandom who had been going to local events, he thought it made sense to go to *Prisoner* conventions in Portmeirion, and he went with his friends starting in the mid-1980s. There, he was impressed by the varied nature of the fandom, consisting of not just the kinds of science fiction fans he knew in Northern Ireland but people from many different walks of life, and by Portmeirion itself. Being able to discuss and play with *The Prisoner* in its original setting was something that Michael greatly enjoyed (he and several of his *Prisoner* friends even did a *Prisoner*-themed role-playing game in Portmeirion at several conventions), and he kept going to PortmeiriCon, as well as other *Prisoner* events, into the early 2000s. Within this society, he made good friends and even met his now-wife. Like many fans of *The Prisoner* who visited Portmeirion, he also got interested in Sir Clough Williams-Ellis's life and work, collecting old Portmeirion guidebooks and reading Williams-Ellis's books.

During the schism of Six of One in the early 2000s (discussed later in this chapter) he left the society and no longer attends PortmeiriCon, but he still keeps in contact with the friends he made while part of the group. He sees them as connecting him to both his fandom of *The Prisoner*, which he is now less involved with than when he was younger, and Portmeirion itself, which he visits about once a year for the "non-convention" of friends who left Six of One but still want to meet up yearly. He also keeps up with information and news about *The Prisoner* and Portmeirion through Facebook, where he is a member of some Prisoner-focused groups. Today, he considers *The Prisoner* as one of his many hobbies and interests, which include Neolithic history and the book *The Owl Service*, and one that has meant a lot to him over the years, even if it doesn't have the same prominence that it did when he was younger. He has even showed *The Prisoner* to his son, who at least seemed to enjoy it.

The Experience and Role of Portmeirion for *Prisoner* Fans

Village Activities: Ritual and Repetition in Portmeirion

Portmeirion is not the most accessible place. Located in the small community of Penrhyndeudraeth in North Wales and nestled on the cliffs of the estuary of the River Dwyryd, it is a 4.5-hour drive from London, 2.5 hours from Manchester, and 3.5 hours from Cardiff. From abroad, the most direct route entails a 4.5-hour train ride from the Birmingham airport to the Minffordd station approximately a mile away from Portmeirion, and then a wander or courtesy car through a residential area before entering the estate's woodlands and descending to the Village proper. Separated from the more populated areas of Great Britain and the small Welsh towns surrounding it, it is a place that must be known about to be found. Once it is entered, it feels like another world entirely.

Despite its isolation, fans of *The Prisoner* have been visiting Portmeirion since its first reveal as the filming location of the Village. Well known as the main filming location, it is considered an important place for fans of the show to visit. For the fans interviewed here, this desire—to see Portmeirion for themselves and have a physical encounter with this environment—was what brought them to Portmeirion in the first place. However, after the first visit, they ended up returning, often on a regular basis. Many have grown to love this place quite deeply:

> [I feel toward Portmeirion] just a great sense of love and gratitude. I find it . . . it was exactly what I needed at all different times I needed it in my life. So it was sort of quite magical and inspirational and exciting to me as a young teenager, when I was in my mid-twenties and kind of unhappy and sort of stressed it was a very cozy place to escape to, it gave me somewhere to write and to think and to be and to feel like I could completely be myself. You know, it's . . . yeah, it's been extremely important to me. And I do feel that I do have this very, very strong relationship with the Village, you know, I have sat on every bench, I've walked in every bit of it, I've stayed in most of the buildings . . . I know it really well. Yeah. It's really important. (Katy, 33, English)

That fans like Katy visit Portmeirion fairly frequently means that they have developed familiar routines within and around it. In returning to see these familiar faces and places, fans also establish practices around Portmeirion that create the sort of pathways that Tuan and Seamon see as the

foundation of a sense of place. These practices can be both formal and ritualized, such as many of the practices involved in the yearly PortmeiriCon
fan convention that I will describe, and more informal practices through
the village and its affiliated landscape. They create a sense of familiarity and
attachment to Portmeirion beyond, but not entirely apart from, its depiction
on screen, giving it continuity as a site of *Prisoner* fandom and a meaningful
location in the fans' lives.

For most of the fans interviewed, PortmeiriCon has been important for
this relationship, either currently or in the past. Fan conventions are one
of the oldest fan traditions, drawing fans to specific locations to congregate
and celebrate their objects of fandom. Ranging from large, multifandom
events like San Diego Comic-Con to smaller, more focused events like PortmeiriCon, they have long provided a way for fans to meet others interested
in the same things and to provide a physicality to what would otherwise be
more ephemeral relationships, either with the objects of fandom themselves
or other fans.

As places, however, fan conventions are seen somewhat differently than
more auratic "pilgrimage" locations. Conventions are typically held in what
Augé refers to as "non-places,"[22] hotel rooms or convention centers that can
accommodate the crowds but without distinctiveness on their own, because,
generally, places that have a stronger association with particular fan texts
do not also have the capacity or desire to host conventions. Rodman stresses
the difference between Graceland and convention spaces in terms of the
way in which the community feels a sense of belonging to the space, stating
that a "*Star Trek* convention, for instance, may afford an otherwise diffuse
population of fans the opportunity to congregate as a community, but the
lifespan of that community is brief (a few days at most) and the site in question isn't likely to be one that has any pre-existing (or lasting) association
with either *Trek* or Trekkies. [. . .] the space occupied by that community
isn't theirs in any sense other than that they've rented it for the occasion."[23]
Graceland, and other auratic pilgrimage sites, have an enduring connection
to the object of fandom that can't be displaced, which makes their resulting
fandoms special.

This assertion is challenged by Hills, who stresses that "over time, specific
hotels may come to take on their own histories of convention-hosting such
that the contingency and alien-ness of the hotel space may actually become
a necessary part of a given convention's identity."[24] This can be clearly seen
in Geraghty's exploration of the San Diego Comic-Con, which has grown

to a massive event in which the "non-place" of the San Diego Convention Center has become a very specific place indeed, with significance given to each aspect of its seemingly generic architecture.[25] Additionally, as Porter discusses, the "non-place" quality of the convention site also promotes a focus on the text(s) themselves, and what they are said to represent for fans, rather than on the space.[26] The very genericness of a hotel or convention center means that it is not the place itself that holds the most meaning, but those in it, and this community can be reproduced seemingly anywhere.

Portmeirion, however, does have both a lasting association with *The Prisoner* and the capacity to host groups. To its organizers, having a convention anywhere but Portmeirion did not make sense:

> It was a logical thing to do. You know. If you start a society for anything, you want to have a meeting. We had a film set! We could come to the film set! Where else do you want to go? You know. If you're a fan of *Star Trek* or *Doctor Who*, you don't do anything. (Roy, 69, English)

Despite its being less accessible than the hotels and convention centers that are common spaces for fan conventions, its symbolic power as "the place" of the show meant that other locations were not options. While *Star Trek* is placeless, set in outer space in the far future and filmed in inaccessible studios, *The Prisoner* is distinctly placed in Portmeirion. Therefore, the fandom must be placed there as well. This is as true for the convention organizers today as it was in 1977:

> There was one year when, I remember now, we couldn't do a convention in Portmeirion for some reason. And there were talks about perhaps in the future we should have them somewhere else, and everybody said, well no! You can't do that! Portmeirion is *The Prisoner*. If you're going to have a convention, it's got to be in Portmeirion. And that's that, really. You know. A proper convention, I mean you can have another event, but an actual convention, a full weekend affair, it's got to be Portmeirion, isn't it? (Angie, 55, English)

Having the convention elsewhere was never a serious option, even as different groups of fans took over the organization of the convention and the relationship with the Portmeirion management fluctuated. Portmeirion therefore combines both of the main ways in which fandom relates to place—it is where the show "actually" happened, where fans can momentarily bridge the gap between imagination and reality, and where they can gather to discuss *The Prisoner* and meet friends who are also enthusiastic about the series.

As a long-running fan convention, the activities and practices of PortmeiriCon are particularly visible as ritualized and repeated practices within Portmeirion. Quizzes about the series, discussions, interviews and talks with special guests involved with the production of the show, episode screenings, and social events are well-established and expected parts of the schedule. These types of events, which adapt well to any type of enclosed space, are part of the makeup of most fan conventions, providing a focal point and structure for the weekend and the discussion of the text(s) that the fans have gathered to celebrate.

In addition, PortmeiriCon also features a number of reenactments from *The Prisoner*. Reenactments and dressing in media-related costumes are standard practices at both filming locations and conventions, as I discuss in the previous chapter, but performing scenes from *The Prisoner* in Portmeirion with the accompaniment of fellow fans has a particular resonance that has made them cornerstones of the PortmeiriCon experience. Each convention features a number of smaller scene reenactments, as well as large, public reenactments of two iconic scenes—the "human chess game" from the episode "Checkmate" and the "election parade" from the episode "Free for All." These are performed at the spot of their filming at least once every convention since the 1980s, with the 2016 PortmeiriCon holding each twice.

Unlike the other convention activities, which require payment and membership in Six of One, these large reenactments are open to day visitors to Portmeirion. Attendees are also encouraged to participate as part of the crowd, with costumes and props handed out to whomever wishes to utilize them. They are often playful, joyous experiences:

> You can just join in, and be as if you are taking part of *The Prisoner*, the filming of *The Prisoner*, but you know. It's almost like you're there and you're part of it and you can have fun. They're fun things. (Angie, 55, English)

For Angie, the ability to join in as a non-actor from her very first convention not only made her feel that she was taking part in the series, but added a sense of joy and fun to the event and immersed her into the fandom, where she is now one of the convention organizers. Participants frequently make jokes as they go through the scenes, and there is a sense of comfortable familiarity and playfulness to the proceedings. They are seen as a way to celebrate the show and the convention through participation, rather than solemnity, similar to the reenactments done at *Game of Thrones* sites. As we see again here, play is an important way of interacting with filming locations.

That they are playful does not, however, mean that they are not taken seriously. This is clear both in the amount of effort put into their organization, compared with the more impromptu reenactments discussed in the other chapters in this book, with speaking roles assigned well ahead of time, scripts sent out, and rehearsals done, and in the feelings of those who are trusted to carry out these roles. Liza was proud of how she worked herself into a position at a previous edition of PortmeiriCon in which she could fulfill a dream—performing in the election parade as Number 6:

> But then as Number 6 I got onto the Stone Boat and I thought "here I am where Patrick McGoohan stood" with this . . . and looked out at all these people, this sea of people and was like "vote for me!" and they were calling back to me. They acknowledged me, they were like "Number 6! You're Number 6!" And it was just like this whole feeling of, of all of us together felt this synergy. This whole kind of "here we are creating it, re-creating it, respecting it!" We are walking in their footsteps and kind of, not in a way to say . . . just to say "this is a moment that we are acknowledging it and doing it together." (Liza, 40, English)

Performing the role in its setting and having it acknowledged by her fellow fans gave her a sense of powerful connection with a character she had long identified with. While it was certainly fun to do, the real joy came out of experiencing this moment. Doing these reenactments in the environment of their filming, in a clearly organized way, sets the reenactments at Portmeirion apart from others. That fans have been doing these reenactments for a long time also gives them a sense of continuity. Liza not only got to experience "being" McGoohan but did so in the footsteps of fans who came before. Maintaining this connection with the past is important for the fans who continue to go to PortmeiriCon.

This is not to say that every fan wishes to be involved with the reenactments. Some fans, particularly those that no longer attend PortmeiriCon, find the reenactments uninteresting or even a bit ludicrous and much prefer other activities, such as discussing the show's themes and spending time with friends, alongside enjoying Portmeirion's qualities as a visual reminder of *The Prisoner* and as a holiday destination in its own right. Even fans who appreciate the reenactments frequently do other things while in Portmeirion, particularly if they also visit outside of the convention. While these other activities are less formalized than the reenactments, that they are done on a regular basis still contributes to the sense of continuity with their fandom and with Portmeirion itself. Geoff, who no longer attends PortmeiriCon but now lives in the area and visits frequently, describes his visits:

> I normally go for a wander through the woods because of the beautiful woodlands they have there. At some point we'll go to the hotel and have something to eat or have a drink or whatever. [. . .] Me and my wife go there just the two of us a few times a year, there are little places we like to go and sit down and have a think and just sit and chat or whatever. We have our favorite little spots in the village. (Geoff, 42, English)

As a frequent visitor to Portmeirion, he has established his own activities and pathways through it, ones that are repeated as he continues to visit and think about it. Seamon states that "regularity and variety mark the place ballet,"[27] and this is what Geoff demonstrates in his interactions with Portmeirion—he has established regular rhythms within Portmeirion, while still treating it as something special. Other fans who visit outside of the convention express similar habits. Wandering, seeing familiar but spectacular views, and meeting with friends at places where they have long met with friends all create a strong relationship with Portmeirion.

Regular visitors like Geoff show considerable familiarity with Portmeirion. They know and discuss the names of each building and area among themselves and debate changes, both actual and potential, with considerable investment, as well as showing interest in Williams-Ellis's other works. Some check the Portmeirion webcams daily, in order to virtually return to Portmeirion and see their favorite places when they are physically unable to. It is a place that matters to them. As Michael notes:

> To us, I can't speak for everybody, but to the people I associate with in *Prisoner* fandom, almost to a man and woman, they are as obsessed or as interested in Portmeirion as they are in *The Prisoner*. (Michael, 52, Northern Irish)

Many *Prisoner* fans interviewed confirmed Michael's observation. Portmeirion is a place that is deeply beloved, and frequent visitors like Michael often find themselves as drawn into its story and history as they were into the show that brought them there. Through the activities of frequent visitation, whether formal or informal, directly connected to *The Prisoner* or not, their affective involvement with Portmeirion deepens. As Roy, one of Six of One's founders and a frequent visitor to Portmeirion, states:

> I'm walking around, and constantly in my mind going, "Yeah, that happened. That scene happened there." I . . . in my ideal world, when I die, I would like to be buried in a shallow grave in the flower bed just below Number 6's house. That is where I would like to spend eternity. Gazing across the amphitheater. (Roy, 69, English)

Roy demonstrates a clear love of both Portmeirion and the program that brought him there, both of which are an important part of his life. His wish would be to dwell there permanently, both in this life and after.

Both Tuan and Seamon saw placemaking through repetitive movement as occurring through everyday activity, but as these fans show, it can also be accomplished through more extraordinary movement—that of continually returning to a special place, removed from everyday activity. Indeed, the isolation of Portmeirion was often brought up as a positive factor. Visiting, either during or outside of the convention, is seen as a time to get away from ordinary concerns and reconnect to the fantasy space. The fans remove themselves from everyday life, but into a familiar and welcoming environment, one that allows them rest, contemplation, and meaningful connections.

Moving in familiar ways is therefore an important part of the relationship that *Prisoner* fans have with Portmeirion, and an important aspect of the homecoming. Through reenacting scenes at the place of their filming, fans not only immerse themselves in and pay tribute to the show that has mattered so much to them, but do the same to the fandom. Informally, continually revisiting and moving around in familiar locations in Portmeirion creates a sense of attachment, adding a personal connection to the symbolic power of being in "the place" where *The Prisoner* happened.

"Welcome to Your Home Away from Home": Portmeirion as Meeting Place

Returns to Portmeirion are not only frequent, but also usually scheduled. This creates a sense of stability for *Prisoner* fans that provides a strong basis for a community. The importance of Portmeirion as the filming location for *The Prisoner* and as a part of the show's mythology means that it has long been a draw for fans. This idea of a "gathering place" for a fandom around a particular site is not unique to *The Prisoner*. Rodman,[28] Aden,[29] and Sandvoss[30] stress the importance of a form of "communitas"[31] around Graceland, the *Field of Dreams* filming site, and Ibiza, respectively, as these locations will always have at least some other fans present, and I recognized a similar phenomenon at the Wizarding World of Harry Potter in the next chapter. Rodman suggests that the constant critical mass of Elvis fans at Graceland "provides enough continuity to foster a lasting, tangible sense of an Elvis-centered community"[32] and reminds the fan that they are not alone in their interest. This is seen more generally as a feature of pilgrimages, which create a "temporary fellowship"[33] of those who are attracted to the site. For fans, particularly those who have interests that are not shared by those around them, being at this important site, with others around for that reason, creates a sense of belonging—that the individual is part of something bigger, and that their interest is not so strange and isolating as it may seem at home.

For fan visitors returning to Portmeirion, it is not only the general sense of a broader community that is important, however, but the relationships with specific other fans. The friendships made via *Prisoner* fandom were considered very important by nearly every interviewee. Compared with the looser communitas that Aden, Rodman, Sandvoss, and I described, the goal is not to be around an abstract sense of community, but to talk with people who share an interest in *The Prisoner*. Meeting other fans was a frequently given response as to why *Prisoner* fans not only visited but returned to Portmeirion. Liza, who has attended PortmeiriCon several times, discusses her decision to first visit Portmeirion:

> I'd never been to a place quite like Portmeirion, and I just really wanted to come and see the place where, for me, the program had such an impact. It was so beautiful and I thought, "I really must come and see this place," and specifically, see other people who enjoy the program. (Liza, 40, English)

A fan of the program on her own beginning in the early 1990s, Liza had found it difficult to find others in her everyday life to discuss the show with. In visiting Portmeirion and the convention, she addressed this lack:

> To be there, years after, and there I was, and the sun was shining, with people who were friendly, people who got me. They got me! When I said to them, "I don't . . . I've got all these questions and there's all these

symbols and meanings that I can read into it and I want to disagree
with you . . ." And it's that whole, to have that fellow feeling of arriving,
in a place. It transformed the place for me. (Liza, 40, English)

In entering this isolated location, Liza encountered not only the "real" places
of *The Prisoner*, but other fans who understood where she was coming from.
Compared with other media "cults," the fandom around *The Prisoner* was
and is smaller. While it might be possible to find other fans of Elvis or *Star
Trek* in one's general vicinity, it was rarer to find a fan of *The Prisoner*, especially before the internet was commonplace. Portmeirion was one of the few
places where it was likely to find one.

Going to Portmeirion to meet fellow fans was expressed as a desire to
have deeper engagement with the program than the fans could find in their
everyday lives. *The Prisoner* is seen by its fans as a particularly innovative,
quality program, one that challenges viewers and requires a higher level of
intellectual engagement to appreciate:

Well, I'm fascinated by television that leaves questions unanswered.
And I like television that stimulates the mind and makes you think.
And I also was taken an awful lot as a child by the idea of the individual
against sort of convention, normality. In fact, all the ideas that *The Prisoner* invokes, I was interested in. (Harry, 54, English)

To its fans, the show not only asks questions in terms of plot—who is Number 6, what is the Village—but also raises questions about society, culture,
and life in general, ones that continue to have resonance throughout their
life course. Many fans wanted to find others who felt the same way about
it. This was not only to discuss *The Prisoner* itself, although that was important, but because they felt that other fans were likely to be interesting people worth meeting. As discussed earlier, fans often identify their object of
fandom with important parts of themselves. This identification can then be
extended to others—a similar interest in *The Prisoner* means that they are
likely to be compatible in other ways. While Sandvoss insisted that most fan
texts are polysemic, allowing "not only for a multiplicity of meaning, but for
any meaning,"[34] within the interviews and the fan-produced texts there is a
reasonably clear sense of what traits *The Prisoner* represents—intelligence, a
questioning of authority and power structures, individualism, an interest in
strange and challenging media—which fans expect to see reflected in themselves as well as others.

For many, this led to both Portmeirion and the *Prisoner* appreciation society, Six of One. The establishment of groups such as Six of One was not uncommon in the media culture of the 1970s and 1980s, and it found its footing among fans wishing to devote more attention to the program and its ideas. As with other fandoms, this was fostered through the aforementioned PortmeiriCon convention, first held in 1977 and repeating nearly every year since (conventions were not held in 1980, 1999, 2000, 2002, 2004, and 2020). This scheduled reason to revisit meant that Portmeirion provided a stable platform to construct a community out of what might have otherwise been a more fluid communitas, more similar to intimate fan gatherings like cruises, which also foster an environment of shared isolation from the everyday and encourage returns.[35] Fans of *The Prisoner* who got involved with Six of One and the convention knew that the people they met there would be likely to return. As time has passed, many of these friendships have become quite close:

> I've still got friends from that very first convention [in 1982] that I'm still friends with now. And as I said, I think that's one thing that links all of us, we've got this love of Portmeirion, *The Prisoner*, Patrick, etc., but we've also got great friendships out of it as well. And it's like, some people you just see that once a year, at the convention, and it just feels as though you've only left them about ten or fifteen minutes or so, and you've only gone for a coffee or something, and you know, there is other ones like my dear friend, who I speak to throughout the year. (Anabel, 54, Hertfordshire)

Anabel is far from the only fan to have made important friendships through the society and the convention, and several have met partners through their attendance. Fans expect to see specific others at Portmeirion whom they consider good friends. Returning is therefore important on two levels: it reconnects individual fans with their fandom of *The Prisoner* and with the relationships they have made through this fandom.

Relationships with fellow fans can be built and maintained in other ways, particularly as communication technology has improved. Smaller, local groups of *Prisoner* fans have also always been a factor in its fandom. However, meeting locally is considered not the same as meeting in Portmeirion:

> No comparison! All we did was sort of sit there and talk about, they would reminisce about conventions they had been to, they'd be telling

me all about it, and then we'd be sitting there swapping *Prisoner* in-jokes and just drinking and having fun, really. Having a few beers and saying a few silly things. Which doesn't at all compare to a convention of course, which is a completely different thing. (Angie, 55, English)

While Angie enjoyed the monthly get-togethers with her local group in the late 1990s, both she and the other fans thought of them differently than the meetings in Portmeirion. Being in Portmeirion—the site of *The Prisoner*, the site of the fandom's history, and physically removed from "everyday" life—added a specialness and uniqueness to the occasion, even if they were simply chatting and sharing a drink or meal as they might elsewhere.

This double significance means that even if fans are no longer willing to attend the convention, they still wish to visit Portmeirion with other fans. A serious schism in Six of One in the early 2000s, arising from personal conflicts and bad behavior as well as disagreements about where the society and the convention were heading, led to many leaving it and therefore no longer attending the convention. However, this did not mean that they were willing to give up regular visits to Portmeirion or the relationships engendered there:

> There is not another place like it, so if you want to spend some time with your friends, which is a great thing to do anyway—I don't know anyone who doesn't like spending time with their friends—what better place to spend it at than Portmeirion? (Geoff, 42, English)
>
> Friends. I would rarely go back there if it wasn't for the people there who I was going to meet up with. [. . .] But then I do take time to walk around Portmeirion itself. We have this almost called NonCon, nonconvention, there's a group of friends who go every August. (Michael, 52, Northern Irish)

While neither Michael nor Geoff wants anything to do with Six of One, they still wish to visit Portmeirion and spend time there with friends they've made on previous visits. Both men were fairly young when they started attending *Prisoner* events, and being a fan, while not their only interest, as they were both keen to stress, is nonetheless a key part of their self-conception that they do not wish to lose. In continuing to visit Portmeirion with their friends, and in continuing to think of Portmeirion as a special place, they maintain this connection even without the structures of organized fandom. Without the formatting of the convention, they can also tailor their *Prisoner* experience the way they want (neither enjoys reenactments, for example).

Continuing to make time to do this on a regular basis maintains their identity as *Prisoner* fans and keeps the friendships secure as well.

For those that have continued to visit Portmeirion over the decades, this sense of fan community is maintained as much in the relationships with others as it is in continuing to watch and discuss *The Prisoner* (and, for some, more so). As Massey suggests, places can be "conceptualized in terms of the social interactions they tie together."[36] Portmeirion is therefore conceptualized for fans as the tying-together of their own personal interactions with the text, described by Williams as a "pure relationship"[37] that the fan builds out of the satisfaction the text gives them, and their interactions with fellow fans, both of which combine to make Portmeirion the special place that it is. Through continually returning over long periods, fans build and deepen friendships, which cement the importance of Portmeirion in their lives.

Many Happy Returns: Portmeirion as Permanence

That fans can expect to be able to return to Portmeirion on a yearly—or more frequent—basis also points to the importance of the longevity of Portmeirion itself. One of the more striking features of *Prisoner* fandom is that it has almost always been what Williams refers to as "post-object": a fandom that exists after the object (the television show) has ended.[38] Williams discusses how being a fan can give a sense of "ontological security" by providing a stable sense of self-identity through both the text itself and the fan community. The text and the fandom make up a part of the fan's life, providing tools to create a self-narrative and a stable sense of who they are. However, "individuals may experience threats to their ontological security through the demise of, or loss of interest in, a fan object or through the failure of fan community."[39] As the fan object stops airing, and the fan communities drift away as a result of the lack of new material, fans might feel a considerable loss as this part of their self-identity ceases to be.

While some of the interviewees watched *The Prisoner* in its first run in the late 1960s, most of them came to the show when it was already a finished, completed work, either through rescreenings or through home media consumption. The structures of its organized fandom, such as Six of One and PortmeiriCon, all came to be in *The Prisoner*'s post-object phase. While spurred by the rescreenings that gave it the rhythms of a current show, *Prisoner* fandom has mostly existed as a fandom without new texts. After the failure of the 2009 remake, fans of the series tend to see the potential of more *Prisoner* texts as a threat to the ontological security of their fandom than a promise of the continuation of it, as it would challenge their conception of themselves as fans of a truly great television show. This puts *The Prisoner* in contrast to other prominent cult texts, which have frequent injections of new material that keep them from fading away from public and fan consciousness.

What it does have is Portmeirion. The ability to gather there on a regular basis gives continuity to meeting up and discussing the program in a way that many other post-object fandoms do not have. Some even suggest that without Portmeirion, the fandom would not still exist:

> I think it's essential. Absolutely essential. It's really essential for the *Prisoner* group. I think it keeps it all together, because it's a meeting point where we come to, and it's beautiful in its own right. There's plenty of time to talk and go off, so everyone enjoys it. It's better than meeting in some place totally devoid of it. (Lily, 71, English)
>
> If it weren't for Portmeirion, I think it would be very hard for the fandom to continue now. Because . . . there isn't the screening anymore. The last time I think it was screened was like ten, twenty years or something like that. So, I think, it would be very hard, especially now, for people to be able to gather in an anonymous place like Birmingham or anything like that. (Liza, 40, English)

For some fans, without the ability to visit Portmeirion and reconnect with their fandom by doing so, they would undoubtedly drift away from it in favor of more visible fandoms and activities. Additionally, the schism in Six of One was very personal and continues to be painful, and could have destroyed the organized fandom of *The Prisoner*. Yet the appeal of Portmeirion itself, the regular cycle of conventions or other gatherings, and the enjoyment that fans have in attending these keeps some form of the fandom, and the identity of being a *Prisoner* fan, secure despite serious disagreements and the passage of time.

There is also, however, another important factor regarding Portmeirion:

> But of course, the thing about *The Prisoner* is that so much of it was filmed in Portmeirion, which is a real place. So it is a living film set, if you like. Not unique in that respect, there's a lot of films and TV shows where you can visit the locations, but Portmeirion, that *The Prisoner* is fairly unique in that where it was filmed is pretty much exactly the same now as it was then. (Geoff, 42, English)

As Geoff notes, Portmeirion is much the same as it was during the filming of *The Prisoner*. While there have certainly been changes in the more than fifty years since the program's filming, they are not considered especially significant to the overall look and atmosphere of the village. The important buildings and other locations used in *The Prisoner*'s filming are all still there, and, for the most part, still look as they did, with no significant alterations or additions. It is still run by members of the Williams-Ellis family, all buildings have historical listed status, and it is set in a conservation area and owned by a charitable trust that ensures it cannot be bought or sold. These markers of stability create a sense that it will always exist in the way it always has, and indeed, did when *The Prisoner* was filmed.

Portmeirion's own stability and sense of permanence contributes an important sense of ontological security to *Prisoner* fans. It is, after all, seen as an incredibly important part of the series:

> [Portmeirion is] a character in *The Prisoner*. It is one of the central char-acters of *The Prisoner*, I'd describe it as that. (Michael, 52, Northern Irish)
> *The Prisoner* and Portmeirion, you can't really separate them. They are bound together. [. . .] And [*The Prisoner*] was obviously born of Port-meirion, because Patrick McGoohan visiting Portmeirion and did some filming there when he was doing *Danger Man*. And he thought at the time, this is the place, I'm going to come back here, we're going to use this place. (Angie, 55, English)

Portmeirion is important to *Prisoner* fandom on several levels—as the inspiration for McGoohan to create the series, as its filming location, and as the gathering point of the fandom throughout its history. The fans interviewed see Portmeirion as an integral part of *The Prisoner*, without which it would be a very different (and potentially less interesting) show, and they use Portmeirion as a way to understand *The Prisoner* and the man who created it, which I have elsewhere called an "unintentional paratext,"[40] something that was not intentionally designed to supplement the show but that fans use as a paratext anyway. The link between the two is very strong.

That this connection, and Portmeirion itself, has endured over the years lends a sense of stability to Prisoner fans, as summarized well by Paul:

> As long as Portmeirion exists, in a sense *The Prisoner* will exist. So you know in that respect everybody [who] is a close fan, wants Portmei-rion to sort of remain. In a sense if you lost Portmeirion you'd lose *The*

Prisoner. So I think in that respect, you know it's a . . . how do you say that . . . it's a much-loved relative. (Paul, 54, English)

Because the association is so strong for fans of *The Prisoner*, the continuing existence of Portmeirion means that, on some level, *The Prisoner* endures. While the general public may or may not be interested in *The Prisoner*, and new material is not forthcoming, Portmeirion's continued existence means that there is still some level of ontological security to their fandom. As Cresswell states, "the very materiality of a place means that memory is not abandoned to the vagaries of mental processes and is instead inscribed in the landscape."[41] In its permanence, Portmeirion remains as a sort of "safe vault" of their memories of *The Prisoner* and its fandom. This is an important element in the fans' place attachment to Portmeirion. Through revisiting, fans have gained a deep attachment to Portmeirion that is informed by, although not always strictly part of, *The Prisoner* itself. However, that it is such an important site for *The Prisoner* means that it is also special for that reason—it is a constant and permanent memory of the show.

In recent years, Portmeirion itself has shown a renewed interest in *The Prisoner*, offering a *Prisoner*-themed high tea during PortmeiriCon, running *Prisoner* tours, and increasingly using the Number 6 branding within the village. For some fans, this is a welcome acknowledgement of the importance that *The Prisoner* has always had to Portmeirion and their own role in maintaining this link:

I think if they searched their hearts, they'd probably think actually, you know, *The Prisoner* put Portmeirion on the globe, on the map. And it certainly created a global attraction. [. . .] I'd think as a very low estimate, that 50 percent of people who go to Portmeirion go there for the *Prisoner* connection. (Mark, 62, English)

That PortmeiriCon, which Mark helps to run alongside his partner Angie (whom he met through Six of One), and its attendees have kept this connection running throughout the years is a source of pride. Mark and Angie appreciate the good relationship that they now have with the Portmeirion management and enjoy seeing their once-maligned fandom valued.

For others, though, this renewed interest is less welcome:

I think it's too obvious. I think there is a subtlety about Portmeirion. And that subtlety is sort of diluted by actually making a crude reference to

The Prisoner. Where I think they should be saying, yes, okay, *The Prisoner* was filmed here for one month, but eh, you know Portmeirion and Clough's dream of the village was way long before that and way after. And you know, it's just one atom of the village's celebrity. And the people will make this association anyway, with *The Prisoner*. (Paul, 54, English)

For Paul, having too many clear references to *The Prisoner* risks overshadowing other aspects of Portmeirion, as, for example, *The Andy Griffith Show* overshadowed the "real" Mount Airy in Alderman, Benjamin, and Schneider's study of the town.[42] There is a security to having Portmeirion exist before and outside of *The Prisoner* that would be threatened by Portmeirion's stressing the bond too heavily, as it would mean that Portmeirion might rely too heavily on the vagaries of popular culture for its survival. Paul would rather that Portmeirion focus on its other qualities, as they better fit the image of Portmeirion as a special place worth preserving, with the filming of *The Prisoner* as one of its many attributes. He would not like to see them crowded out by "tributes" that fail to match the quality and taste level that *The Prisoner* has for him. Portmeirion's uniqueness is already enough of a fitting reminder of *The Prisoner*.

Essentially, what Portmeirion ultimately provides for many fans of *The Prisoner* is the permanence of place. In its physicality, monumentality, and lack of change, Portmeirion can function as "safe vault" for *The Prisoner* and its fandom. Rather than existing only as ephemeral media or personal memories, *The Prisoner* is instead tied to an enduring physical place—Portmeirion.

Conclusion

What is the potential long-term role and significance of place in fandom? In this chapter I have answered this question by examining long-term fans of *The Prisoner* and their relationship with its filming location of Portmeirion. Many of these fans have been visiting Portmeirion regularly for decades, returning at least once a year to reconnect with their friends and their fandom. Compared with the single trips that have been the focus of much of the discussion of fandom and place, these return visits show a different role for both individual fans and wider fan communities. I suggest that these more frequent trips should be called "fan homecoming"—a return to a familiar and beloved place associated with the object of fandom, one that fans have gained a strong attachment to over their years of visiting. By exploring how this attachment is built, this paper broadens the current discussion on film and fan tourism by highlighting its long-term potential and implications.

Through fans' continual movement to and through Portmeirion, sustained through the yearly PortmeiriCon convention as well as personal visits, Portmeirion has become a deeply felt place. It is important for both the individual fans and the fandom as a collective, sustaining their attachment to *The Prisoner* and their fellow fans as the years have gone on. By continually gathering and repeating certain actions, such as scene reenactments and fan discussions, fans pay tribute not only to the show, but to the fans that have come before and their own past selves. This repetition also creates an attachment to Portmeirion itself as not only the site of *The Prisoner*'s filming, but a place they feel "at home" in—one where they can escape from everyday concerns, reaffirm the importance that *The Prisoner* has for them, and reconnect with their friends and their memories.

Because many of the visits to Portmeirion are scheduled, Portmeirion becomes a place in which a fan community can be constructed and maintained. The yearly schedule of conventions and other gatherings provides a stable basis to build a more intimate fan community, as it is expected that other fans of *The Prisoner* will be there at the same time. It has become the center of the fandom: to truly connect to it, one must visit Portmeirion, and preferably in the company of other fans. It could also be argued that the focus on Portmeirion as the "only" place of the fandom has limited *Prisoner* fandom to those who can access it, and those who have been going there long enough to understand the way in which the fandom operates in the space. Compared with a more "placeless" fandom that can be reproduced everywhere, fandom of *The Prisoner* is considered to be most authentically experienced in Portmeirion. That many of the returnees have been going there for decades also creates a somewhat insular community, focused as much on itself and its history as on *The Prisoner*. However, it is one that undoubtedly means a great deal to its participants, and one that has kept itself and its vision of the show alive and relevant over the decades.

Portmeirion also provides an important sense of ontological security to fans of *The Prisoner*. In its physicality and relative stasis it provides a tangible connection to the show that is expected to endure well into the future. That both the show and the fandom are tied so heavily to Portmeirion means that the fans feel that both will endure as well. Compared with other post-object fandoms that lack such a physical link, *The Prisoner*, it is felt, will never "really" die. It will always have a permanent memorial in Portmeirion.

What is particularly interesting is the continued strength of the relationship between fans of *The Prisoner* and Portmeirion. It is unsurprising that Portmeirion still attracts some fans of *The Prisoner*. The show continues to be a major touchstone in television history, and with both nostalgia for the past and new television distribution technologies remaining prominent factors in contemporary cultural life, that its filming location would continue to attract tourists is to be expected (although certainly worth mentioning in response to the question of whether film tourism can ever be more than "in the moment"). That many of the same fans have been visiting and revisiting Portmeirion for decades, however, is intriguing. They were drawn to it because of both *The Prisoner* itself and the PortmeiriCon conventions, but the connection goes deeper. This deep bond between some of the fans and the filming location has been one of the more surprising, and interesting, aspects of this case, evident to me from my earliest interviews and continuing throughout. While some fans interviewed were less attached than others, that enough of the bonds were so strong, and that all fans expressed a clear fondness for Portmeirion the place, means that it should be highlighted in discussion of what place can be for fans and fandoms.

The tight link between Portmeirion and *The Prisoner* fandom is perhaps a unique one. Few locations provide the same combination of filming location, convention space, and longevity that Portmeirion has. However, this more extreme example illustrates the potential of how place and long-term fandom can interact. The idea of the homecoming developed here draws attention to important aspects of the relationship between fandom and place that have been previously overlooked, particularly the impact of longevity and frequent rather than single visits. Media texts increasingly live on in cultural memory, and fandom is considered part of long-term identity formation; therefore it is worth considering how having a place to go can shape a fandom over a long period. As I show in the next chapter, this relationship is also something that can be created from the ground up.

The Wizarding World of Harry Potter

Immersion, Authenticity, and the Theme Park as Social Space

"That's what [the Wizarding World of Harry Potter] is about. Living it. Being transported to that world and living a day, or a night, there. And that will be completely and truly MAGICAL," reads a Tumblr post by user niallgirlalmighty, as part of a description of the Wizarding World of Harry Potter in Universal Studios Orlando and Universal Studios Islands of Adventure.[1] This post joins many similar ones by fans of the Harry Potter series, who have been flooding to the parks since the first Wizarding World opened in 2010. The parks feature not only rides, some in 3-D, but complete re-creations of fictional locations from the Harry Potter series—the village of Hogsmeade, site of the central location of the Hogwarts School of Witchcraft and Wizardry, and the London wizard-only shopping district of Diagon Alley, each with architecture, shops, and restaurants featuring products and imagery from the series. Its success has led Universal Studios to open Wizarding World of Harry Potter areas in both its Osaka and Hollywood parks. The Wizarding World exemplifies a push in the industry toward more immersive theming around known narrative worlds, going beyond rides and souvenirs into full, complete environments promising immersion into a favorite text. As of writing, the Wizarding World of Harry Potter has been joined by two competitors at the Disney parks—Pandora: The World of Avatar and Star Wars: Galaxy's Edge—both hoping to re-create the Wizarding World's success by providing an immersive themed environment to parkgoers.

Visiting theme parks has been a popular pastime since they started appearing in the mid-1950s, yet there is still little understanding of what makes them so. Research has traditionally focused on a critique of their form, particularly from a postmodern perspective that focuses on the role of simulation in an image-focused society, overlooking the visitors themselves and how they make meaning out of such simulated environments, although this is starting to change. This means there is only a limited understanding of why theme parks actually appeal to the millions who enjoy them. There is also little understanding of how these parks, and the experience at them, connects to other places associated with films or television shows. As Davis[2] and Koren-kuik[3] suggest, theme parks are built on an understanding of outside narratives—they are a way to connect with them on a spatial level. This means, essentially, that they are places of film tourism. It is from this perspective that I began this research.

However, compared with the cases profiled in the previous chapters, the Wizarding World of Harry Potter is a very different kind of environment. It is not a site of filming, but rather, of re-creating the environment of filming. Such re-created spaces have been dismissed in other studies of film tourism as inauthentic, as they lack the aura granted by filming, but the success of the Wizarding World, and the way it has been embraced by fans, suggests that it is time to reconsider this perspective. Therefore, in this chapter, I investigate how visitors interpret this simulated environment, and what leads them to embrace it in the way that niallgirlalmighty and others have embraced the Wizarding World of Harry Potter. By doing so, I explore the complex interactions of fandom, commerce, and physical space in the twenty-first century, and investigate how a sense of "being there" can be created from the ground up.

I suggest that the success of the Wizarding World of Harry Potter can be credited to its understanding as an authentic adaptation of the Harry Potter story-world, a place of spatial transmedia[4] where Harry Potter fans can employ their ironic imagination[5] and experience the story-world in an embodied manner. The implications of these findings will be explored by first discussing the nature of theme parks as a form or medium, one that potentially enhances the immersive aspect of contemporary storytelling. Following this, I analyze the interviews in the context of my observations in the park, presenting a new understanding of how simulated places are understood, experienced, and appreciated today.

Understanding the Wizarding World

In most intellectual traditions, the theme park is seen negatively. Eco is discomfited by the idea that "Disneyland tells us that technology can give us more reality than nature can,"[6] while Sorkin suggests that "a trip to Disneyland substitutes for a trip to Norway or Japan" and in choosing Disney over these locations the tourist "has preferred the simulation to the reality."[7] The theme park from this perspective dangerously destabilizes the separation between reality and unreality, especially for unsophisticated tourists who seemingly "prefer" the fake. They are also seen as generic, lowest-common-denominator entertainments that fail to fully represent what they are re-creating in order to have mass appeal. Their success points to a postmodern preference for simulation, safety, and entertainment over the "real" experience of landscapes and environments, and the flattening of everything in postmodern life to mass-produced images.

Even those who take a positive view toward popular culture frequently dismiss theme parks. Aden contrasted the experience at the filming location for the film *Field of Dreams* to that of being at an amusement or theme park, stating that "fans treat the field as something special rather than as an ordinary amusement park-type site."[8] The lack of entrance fee and ability to wander create, to Aden, "a special place full of communitas not found anywhere in the mundane, consumerist habitas."[9] Theme parks, with their boundaries and explicit commercial purpose, are contrasted with this purer, noncommoditized location. Similarly, Hills's concept of "cult geography" is defined as "fan attachment to non-commodified space, or at the very least, to space/place which has been indirectly or unintentionally commodified so that the fan's experience of this space is not commercially constructed."[10] The commodified space of the theme park is compared with the noncommercial, individual tour of *X-Files* filming locations in Vancouver, considered a true space of cult geography as the experience is constructed through the fan's meaning-making processes rather than the media or tourist industry. It is this line of criticism that Gilbert's autoethnographic critique of the Wizarding World of Harry Potter internalizes, as she determines that her previous, self-directed fan practices were more "authentic" than her visit to the Wizarding World, which was only for "leisure and tourist consumption."[11] Both Aden's *Field of Dreams* site and Hills's Vancouver also point to a valorizing of "real" locations connected to popular culture. As is discussed in regard to film tourism, the actual place of filming is thought to have a powerful aura that fans seek out in order to truly connect with a favorite narrative.

At the heart of all these critiques is the idea that theme parks are inauthentic spaces—either in that, as artificial landscapes, they substitute for "real" experiences of place or, because of their lack of connection to actual filming and/or their commercial purpose and design, they are unsuitable for authentic engagement with favorite texts or fandoms.

However, this is starting to change. Lukas argues that themed environments have their own form of authenticity—one based on their multisensory aspects. A themed environment becomes authentic when it "is sensory available."[12] Those visiting know it is a simulation, but it becomes an authentic one when it feels correct on all sensory levels. Similarly, Clavé stresses that theme parks should be thought of as "cultural creations equivalent to a painting, a photograph or a film."[13] This is not to say that they aren't commercial, corporate, and geared toward consumption—but that they should also be evaluated as creative productions. A theme park is "a place of fiction that bases its existence on the materialization of a fantastic narration through shapes, volumes and performances."[14] Rather than examples of society's preference for sanitized versions of reality, theme parks are specific places in which fantasies, mythologies, and cultural icons can be enacted and played with. This engagement is heavily visual, but by no means exclusively so, as theme parks and rides also present a multisensory experience not utilized in other media. As with other art forms, they are meant to be an interpretation of a story, and are no more or less challenging to the idea of reality.

Physical separation is a key characteristic that makes them effective as a cultural form: "the 'vocation' of parks is to be worlds apart."[15] Combined with the fantastic narration of theming, the visitor is encouraged to experience them as not the same place as outside. The park functions as an artwork, and "as a real object inscribed in space and time, the work of art is in the world, but as a virtual object that creates its own space and time, it is not of the world."[16] Within the enclosed space, the visitor is encouraged to engage with the fictionality of the theme(s), to pretend and imagine on a bodily level while inside. That theme parks can be understood as artworks means, as Williams[17] explores, that there can be such a thing as theme park fandom, made of those who see and appreciate the art and craft of the theme park and become fans of the multisensory environments they create. Williams refers to theme parks as "spatial transmedia," "moments of narrative extension and world-building that take place within specific rooted locations,"[18] and shows how fans have developed a sophisticated, complex participatory fan culture around these spaces. If theme parks are a medium, they can therefore be judged, evaluated, and appreciated for what they are—a way of engaging with narratives on a spatial, multisensory level, rooted in specific locations.

Most of the narratives used are, however, known from outside the park. Davis explains theme parks as a version of "media convergence" before this term was in vogue: a place to provide new interpretations and promotions of the same media text, utilizing different senses in order to provoke interest.[19] As Mitrasinovic explains, this model was pioneered by the Disney parks, which use familiar landscape archetypes such as "small-town America" and combine them with the familiar brands of the Disney media empire.[20] Korenkuik describes the Disney parks specifically as "spatial narratives" that present a different form of engagement with the familiar characters and worlds of Disney, requiring physical presence and corporeal senses.[21] What the park offers is the chance to physically interact with characters and narratives that are well loved, made all the more powerful by the long history of Disney media, its existing cultural meanings, and the interaction of different Disney texts with each other. The Disney model of design and cross-marketing is the archetype on which contemporary theme parks are based,[22] including Universal Studios, which utilizes popular "blockbuster" narratives like *Jurassic Park* and Marvel superheroes in its rides and environments.

Beeton makes a more explicit connection between theme parks and film tourism, noting that Disneyland, and later parks such as Universal Studios, Warner Bros. Movie World, and Fox Studios Australia, were specifically built around film itself as a theme.[23] They often originally focused on offering a form of "backstage access" to film production, drawing on what I in chapter 2 refer to as the "production mode" of imagination, showing how films were made through viewing of either actual production spaces or simulated versions. As Beeton's example of the failure of Fox Studios Australia reveals, however, just showing production in some fashion isn't enough to sustain a park's profitability. It must be linked to something more exciting, in terms of either thrill rides or a direct connection to popular narrative worlds. As Williams shows, this reliance on expansive narratives has become quite complex in recent years. Rides and environments have developed their own creative fandoms and become sources for other Disney media texts like merchandise and films, with the highly successful *Pirates of the Caribbean* series as the exemplar, further developing the narrative world of the rides that fans have speculated about for years[24] and bringing it to new audiences. As both Beeton and Williams note, this in turn has the potential to draw in fans of the films who want to see where it all "began." Alternatively, the more successful "movie parks" like Universal Studios have moved away from the "backstage" model and toward a more hyperdiegetic connection to filmic texts. The Wizarding World of Harry Potter follows in and builds upon this tradition, using the idea of "entering into" the film itself as its starting point.

Compared with other theme parks, including other areas within Universal Studios, the Wizarding World takes a more holistic approach to its theming. It is presented as a complete reconstruction of locations from the Harry

Potter series, rather than an environment that uses elements from it to create a general sense of fantasy. Specifically, it reproduces the village of Hogsmeade and the urban neighborhood of Diagon Alley as they appear in the Harry Potter film series, with a sense of scale that matches the proportions of an English urban environment. In doing so, it comes closer to the idea of simulation as utilized in media and game studies, where it is seen as a key aspect of new media storytelling and artistic technologies. A simulation creates a sense of immersion: feeling, through the medium, that the audience member is part of the artistic or narrative world. According to Huhtamo, "the quest for immersive experience is a cultural topos, which has been activated—and even fabricated—now and again in culturally and ideologically specific circumstances."[25] As I mentioned in chapter 1, this is part of a long tradition in Western art and technology, which frequently seek to create a believable illusion of reality through "artificial" means.

This search for immersive experience is most often applied in theme parks to the simulation ride,[26] but we can also think of the entirety of the theme park experience in this fashion. They are enclosed spaces in which the visitor enters to feel immersed in fiction. Ryan argues that "the Disneyland tourist, beloved scapegoat of cultural critics, deserves credit for the ability to appreciate the art that goes into the production of the fake."[27] By this, Ryan refers to enjoying the sense of immersion in a fictional story-world and the appreciation of the skill that makes it believable. She argues that this is one of the great pleasures of narrative, a "fundamental and timeless dimension of aesthetic experience."[28] It is what gives fiction its enduring place: the framework to imagine potentialities and alternatives, to explore what could be.

Ryan's concept of involvement and immersion in narrative worlds is further elaborated by Saler. Rather than immersion through technology, Saler stresses immersion through imagination. He investigates the broader cultural construction of what he terms "virtual worlds," "acknowledged imaginary spaces that are communally inhabited for prolonged periods of time by rational individuals."[29] To Saler, a story-world becomes virtual when it is adopted and discussed by many individuals, who group together to explore and fill in its details and make it more "real." The story-world becomes immersive because it feels inhabitable—as detailed as the "real world" and shared with others as a sort of imaginary habitus.

It is this experience that Universal Studios reproduces at the Wizarding World through the medium of the theme park, utilizing its characteristics to make physical a process that Saler depicted as cerebral. I argue that in drawing on the park's separation from "everyday life" and its promise of physical interaction with a story-world that is already a "virtual world" in Saler's terms, Universal Studios sets up a situation where experiencing the created, simulated space can potentially have the kind of meaning that has been previously assigned to more auratic film tourism locations. I now focus on how this works in practice.

Methods

Initial fieldwork and participatory observation were conducted at the Wizarding World of Harry Potter in Orlando in late December 2014, the first winter holiday after the opening of Diagon Alley, as this is a peak time for theme park visitation and the highly anticipated expansion attracted many fans, even if they had already visited the park. I conducted participatory observation over three days, spending approximately 10 hours per day in the Harry Potter areas of the two Universal Studios parks (considered collectively here as one park, as fans often do) and primarily experiencing the park as a visitor, but also observing other visitors and their interactions with the park. While privately owned, theme parks are largely treated as public space, and observations were made in an unobtrusive manner without references that could identify individuals or specific social groups. I did not approach the park's management with my plans, as I wanted to maintain my independence from the organization in terms of how I operated and my findings.

I met the majority of interviewees in the Wizarding World at the time of fieldwork and approached them based on visible fan behavior in the park such as attire, reactions to certain attractions, or reciting book or film dialogue. While subjective, such cues serve as a way to identify oneself publicly as a fan to other fans, which is frequently desired at the Wizarding World.[30] After the initial introduction in the park they were contacted via email to further explain the research and set up an interview, in response to which thirteen then agreed to be interviewed. Two interviewees were recruited via the social networking site Tumblr, where fans share posts about the Harry Potter series and visits to the parks, both to reach saturation and to ensure that the study included perspectives from fans involved in the online participatory communities that are emblematic of contemporary fandom. The interviews were conducted via Skype several weeks after the initial trip. As I found in conducting my fieldwork for chapter 2, conducting on-site interviews while participants are on holiday has serious drawbacks, particularly in a place that is as loud and crowded as the Wizarding World of Harry Potter. While interviewing via Skype sacrificed the immediacy and physical presence of an on-site or face-to-face interview, it allowed for greater comfort for the interviewees as they could set their own time and place to be interviewed, away from the crowds and noise of the park itself and without sacrificing any time out of their holiday. This also led to a more reflective attitude, as they had time to process their experience at the park and to put it into the narrative they wished to share.

Interviews lasted approximately thirty to seventy-five minutes and were semi-structured and open-ended. This allowed for a comparison between interviewees on specific aspects of the Wizarding World experience, but also permitted following up on particular themes as they emerged. Questions addressed the existing relationship to the Harry Potter series, the reasons for wanting to visit the Wizarding World, expectations for the visit, activities and emotions while there, standout memories from the park, purchases made while there, and, finally, their feelings about the park at the time of the interview. In total, fifteen visitors were interviewed, three from Canada, one from China, and the rest from the United States, with ages ranging from seventeen to forty-four, with the majority in their twenties and early thirties. Thirteen interviewees were female and two were male. This is somewhat representative of the park experience, although not entirely. I did not seek to interview children, who make up a noticeable portion of the park attendees, and while I did interview one park visitor from China, the proportion of visitors from Southeast Asia was higher than I could find among willing participants. All interviewees were informed of the purpose of the research before their interview and have been given pseudonyms to further protect their anonymity.

Experiencing the Wizarding World

Locating Authenticity in the Theme Park

No movie filming has taken place at the Wizarding World of Harry Potter. It is also far away from the setting of the books, which has its own tourist industry based around the "original" landscapes of the series.[31] Nor does it boast relics from production, as is the case with popular culture exhibits in museums and elsewhere,[32] including at the Warner Bros. studio tour in London.[33] However, most of the interviewees still found visiting to be a powerful experience that connected them with the Harry Potter narrative world:

> It's so fantastic there and it's just, it's like my favorite book come to life, and . . . like, I love this. Like, this is awesome, this is home, like I, I felt like, I had come home to some part of like, the small part of me that

Maggie is twenty-one and from Kansas, where she's a student at a college of education. She has been a fan of the Harry Potter series since second grade. She is involved in online and offline Harry Potter fandom; writing and reading fan fiction; running some Harry Potter discussion groups on Facebook, from which she has met Harry Potter fans from around the world; and attending wizard rock concerts and other Harry Potter events around her home in Kansas. Via these activities, she had heard about the Wizarding World of Harry Potter and that it was a great experience for fans, but she didn't think she would get the opportunity because of the cost. Then her friend invited her as part of her twenty-first birthday celebration, and Maggie got to go to Orlando.

Once there, she felt that the Wizarding World surpassed her expectations. She had only been to local amusement parks and had not had the full theme park experience, so the deep level of theming in its rides and attractions impressed her. More importantly, the attention to detail, and the accuracy of those details based on her knowledge of the series, was more than she expected. It looked and felt like the place she had come to know over many years of caring about the story-world, and she could walk around in it and experience it for herself in the way that she had dreamed about since she was a little girl. She even felt that it was a better Harry Potter experience than the films were—it was more accurate to the books.

In going with her friend, she also had someone to share this experience with. They could go into the shops and sing fan songs with each other without feeling that they were being particularly weird, playing off each other's enthusiasm for the place. They spent three days at Universal Studios, most of which they spent at the Wizarding World sections—taking pictures, riding the rides, and shopping. Maggie bought several souvenirs and was impressed overall by the breadth of the merchandise available—much more than in her local stores, well made and well priced, and with the accuracy that impressed her throughout her trip.

She recalled her visit to the Wizarding World very fondly, joking that she would hole up in Hogwarts and become a squatter in order to never leave. She described her childhood as rough, but she found hope in the series—hope that love would win out, that she could get over loss, that there was hope for her to succeed in her life despite everything that had happened, as Harry had succeeded in his. Visiting was an emotional experience, as she felt that her favorite series had truly come to life and that she could be part of it.

wishes more than anything that I could've entered that world, and that part of me feels more at peace now. (Maggie, 21, American)

Maggie was not alone in discussing the park in terms of "finally" entering the story-world or even being in some way "home." Despite being a theme park in Orlando, it can still invoke an authentic sense of "being there." If we think of the park as a cultural creation or medium, it opens up the potential for a legitimate encounter with the story-world. The park is accepted as a valid adaptation.

This acceptance of the Wizarding World of Harry Potter was credited to a few main factors. For several interviewees, the (perceived) involvement of Harry Potter series author J. K. Rowling was important:

I heard J. K. Rowling actually had a say in everything, saying yes, that can be done, and no, that can't. So I think . . . the fact that she had a hand in bringing her creation to life, and has such high standards, I think that helped. (Circe, 29, American)

Authenticity for these fans is based on the figure of Rowling and her approval of the park. As the series' "brand guardian,"[34] despite some recent high-profile controversies, her approval gives the park a sense of legitimacy. If something is there, then it is approved, and the experience is likely to be a good one if she signed off on it.

Most mentioned, however, was the amount of detail from the series present in the park:

Everything is perfect, it's really like you are there, it's exactly how it is described in the books, and it looks even cooler than it does in the movies, because in the movies—you know what I'm talking about: "Wait, that, what was that back there behind him?" and now you really can look at it, and having been to the actual set in London and then to the Wizarding World in Florida I can say it is so much more detailed, like they really went all out. (Jen, 30, American)

As Jen discusses, the amount and specificity of the details in the Wizarding World were considered crucial to its sense of realism. That it is "exactly described" as it is in the books, visually matches the films, and provides more details about the world than watching the films or even touring the set makes her feel that she is "really" there. Interviewees regularly mentioned the level of detail as one of the things they most enjoyed about the Wizarding World, especially the "smaller" details that they felt might have been easily overlooked. These details coalesced into a sense that this was a realistic depiction of the narrative world, one that held true to the image the fans had formed from years of familiarity. It therefore lived up to the ideal of the theme park—an environment where a cultural fantasy can be engaged with in a multisensory way.

For many, it does this so well that that other theme parks are seen as inferior:

> The little details [. . .] make it feel like a real experience, not like a theme park. (Dan, 33, American)
> The other theme parks are like watching TV about Europe, where the Wizarding World is being in Europe. It's just two different planes of comparison. (James, 44, American)

For Dan and James, the Wizarding World feels more sensorily authentic than other theme parks. Its details—from the signs in shops to the British-style food and drink—create a sense of specificity that in turn fosters a sense of immersion in and authenticity of the story-world, especially when compared with the caricatured theming of other parks.[35] For these fans, it is no longer just a theme park, with all its negative connotations. Instead, it is an accurate adaptation of the Harry Potter story-world.

That the film series is an adaptation as well also provides an interesting comparison to other locations of film tourism, including other locations related to the Harry Potter series. The film set is now a tourist attraction outside of London and presents itself as the "real" place of filming, playing up its connection to the production mode of film tourism[36] (with the occasional foray into hyperdiegetic immersion, particularly for special occasions). Tourists there can see the actual props and sets, learn how the movies were made, and purchase souvenirs as well as some of the series' iconic food and drinks. London also boasts Platform 9¾ at King's Cross train station, where Harry and his friends leave the "ordinary" world to head to Hogwarts every year. As one of the few locations that exist in both our world and the story-world, it is a popular attraction for fans, and has been transformed from a location fans had to find for themselves to a large, stage-managed photo opportunity with "authentic" props and an official store.[37]

However, while these other places can make more claims on traditional authenticity, the Wizarding World of Harry Potter stands up to them in terms of authentic experience and, for some fans, surpasses them. Platform 9¾, as Larsen[38] describes, is a brief and highly scripted experience that has been moved from its "actual" location on the train platforms to a more easily controlled space nearby, removing some of its specific aura. Many fans also remember its prior existence as a more authentically "hidden" spot within the bustle of a functioning train station, when it had shades of the historical mode of film tourism, and so the experience at "Platform 9¾" is somewhat diminished. The studio, while the actual site of filming, has no more claim on being the real Diagon Alley or Hogsmeade than the Wizarding World has. They are both interpretations of a space that first existed in the imagination, and therefore both can serve as "places of the imagination"[39] that link the imaginary world and landscape to ours.

As suggested by Jen, the immersive factor of the Wizarding World is higher than at the studio tour, allowing her to have a closer, more intimate and personal experience with the imagined world of the series. The reproduction of imagined details into physical space gives the Wizarding World a convincing presence:

> They were exactly like they were in the movies. So, you just felt like you were actually there and you're like "oh, Harry and Hermione and Ron were here but they're not here now." (Melissa, 21, Canadian)

This sense of being at the right place at a different time is the ultimate expression of what I in chapter 2 call the "hyperdiegetic" mode of the film tourist imagination—the feeling for the tourist that they are in the narrative space and able to explore it. As it is "exactly" like it is in the movies, this fantasy is realized for Melissa—it is as if she is in the same space where Harry and his friends have been. While the studio tour might be the actual place of filming, its structuring around the production mode of film tourism cedes the hyperdiegetic mode to the Wizarding World, which far surpasses the Platform 9¾ experience in terms of detail and ability to spend time there. The Wizarding World is the place to be immersed in the narrative, as I will discuss, rather than to learn about how it came to be.

The existence of the story-world across multiple media means that the theme park is seen as simply another depiction. Its authenticity is judged on its own character as a medium—its ability to represent the story-world in physical space. Because it is accurate, whether the fan has the books or the films in mind, it feels not only valid, but like good art.

If, as Crouch argues, it is through embodiment—"a process of experiencing, making sense, knowing through practise as a sensual human subject in the world"[40]—that we gain a real sense of a place, then the visitor on some level gains a sense of Diagon Alley or Hogsmeade through visiting the Wizarding World. As an interviewee recalls:

> My daughter said, "I just wanted to bottle the air." Because it smelled of sugar, and not the sickly carnival sugars, but, sugar! And, you know, and this place, you could hear as you passed the shop with the cauldrons you could hear kind of the bubbling and the mandrakes were having a screaming . . . (Daphne, 44, American)

In the multisensory aspects of the theme park, the story-world is engaged with in an embodied and immersive manner. While theme parks are often dismissed as strictly visual, Daphne and her daughter show, echoing Lukas[41] and Williams,[42] that this is not the case. Tastes, smells, sounds, and physical movements that are part of the narrative world are experienced through the park. This experience gives visitors an embodied sense of a story-world that, while familiar, was previously only cerebral or audiovisual. Once the park's simulation is accepted as authentic, the visitor feels as if they are having a genuine encounter with the narrative world.

The layout of the Wizarding World of Harry Potter encourages this. If the theme park is understood as a set-aside place in which to engage with the imagination instead of everyday life, the Wizarding World extends this further by separating the Harry Potter sections from the rest of Universal Studios. It is entered through either a gate (Hogsmeade) or a "London" building façade that hides the "magical" part from view (the newer Diagon Alley), with a dramatic reveal as the visitor fully enters. This physical separation from the rest of the park is similar to the series' separation of Muggle and magical societies and adds another layer of encouragement for the visitors to feel as if they have entered another world. This is an emotional experience for many:

> You just walk in and you see the Alley, it's just . . . it made me want to cry. I was so excited, and so happy. It was just so overwhelming. (Elle, 17, American)

There are many such reports of amazement and tears when visitors step into Diagon Alley for the first time. It is as if their dream, that they have received the letter of admission to Hogwarts that allows them to enter magical society, has come true. Visitors are encouraged to think of themselves as leaving their mundane existence behind them, just as Harry did.

The series means a great deal to its fans. They have long wanted to inhabit the narrative world, in some cases for much of their lives. This existing desire shapes the way they interact with the park and its attractions:

> I just wanted to ride the train. I would have been happy if it was just like, you could see Orlando through the window, but they have, like, a scene that goes by, and that's so cool. But it was just cool just to sit in the compartment and have that experience like going to Hogwarts and then coming back. (Hanna, 26, American)

The series has been important for much of Hanna's life, and she has long dreamed of riding the Hogwarts Express, the train that takes magical students from London's King's Cross station to Hogsmeade. The Wizarding World uses a version of this train to connect the two areas of the park, entered through replicas of the two train stations and featuring short films outside the train's "windows" depicting the journey that the rider is imaginatively embarking on. While the ride version differs from how the train is described in the books and films, it is enough like its description that, combined with Hanna's existing wish to travel to Hogwarts and the immersive environment of the park, it made her feel "like" she was having the experience. Saler refers to this as "ironic imagination," a "double consciousness" that allows the (modern) subject to be emotionally invested in and contemplative about a fictional world, while maintaining the knowledge that it is fictional.[43] While Saler applied this to imagining and discussing a story-world, without the physical component, it is equally applicable here. Those who participate gain a sense of how it would feel if the story-world were not fiction, but they never lose sight of the fact that it is. Rather, they play with the idea that for a brief moment the lines between fiction and reality blur, a pleasurable pretense made all the more so by the effort they feel the park puts into the illusion. That the Wizarding World re-creates story locations that don't otherwise exist also contributes to this exercise of the ironic imagination. The visitor knows that they are not actually in Diagon Alley or Hogsmeade, but there is no more "real" version, and it is a physical experience with all the cultural markers of reality. This makes it a convincing pretense, one that matches the existing imagination with physical sensation, an illustration of how Williams's concept of spatial transmedia creates a meaningful experience for fans.

The train is not the only ride found in the Wizarding World. In addition to the child-friendly roller coaster Flight of the Hippogriff and the thrilling Hagrid's Magical Creatures Motorbike Adventure, both found in the Hogsmeade section, each Wizarding World area boasts a modern simulation ride. In Hogsmeade it is themed around the Hogwarts school, and the one in Diagon Alley is themed around the Gringotts bank, both important locations in the series. Both utilize a combination of screens, 3-D projections, props, and movement to create a sense of kinetic immersion in the ride's story and feature heavily decorated and narrativized areas for waiting in line. Both were praised by the interviewees. However, when discussing their favorite parts of the park, they rarely mentioned the actual experience of riding. Instead, it was the experience of the environment—both the detailed queues of Hogwarts and Gringotts and the outer areas of Diagon Alley and Hogsmeade—that was brought up.

It is in these areas that the Wizarding World most closely matches the "virtual world" of the Harry Potter series that fans already mentally inhabit. While the rides are appreciated, they are too controlled to feel inhabitable. Within the broader environment, however, there is a much stronger sense of agency. Laid out as curving, intersecting city streets, with several potential directions for movement, the Wizarding World is a scripted space rather than a set one.[44] The visitors can take as much time as they want within it and, once past the entrance, wander in whichever direction they choose, actually quite similar to how Aden described the utopian, noncommercial *Field of Dreams* site.[45] Scripted spaces create a sense of narrative centered around the walker, one with the illusion of agency.[46] Compared with how narrative agency is theorized in the context of computer simulations,[47] however, here it is minor, confined to activities like walking, eating, and shopping. Rather than controlling the narrative direction, the visitor to the Wizarding World is cast as an "ordinary" witch or wizard visiting the location. As the narrative world already has meaning and presence, this is enough.

At the time of fieldwork, the Wizarding World was a frequent site of cosplay, dressing up and embodying characters from the story-world, and some visitors create a character for themselves while there:

So, we actually took a few minutes, my daughter and I, and we figured it out. My husband is Teddy, that is why she has the blue hair. And I am the daughter of Bill and Fleur, that is why [son] Ben has the red Weasley hair. (Daphne, 44, American)

At the park I'm usually a Death Eater or a Slytherin student. (Circe, 29, American)

The idea of the park as a place to pretend encouraged this use of cosplay—of fully embracing the idea that you are part of the story-world. A sense of agency in the narrative world encourages the idea of "as if." If the magical world were real, they would do these kinds of things in it. They can (ironically) imagine that they are witches and wizards themselves, and the embodied physical sensations like movement, taste, and smell create an even stronger knowledge of the story-world.

Much of this embodied experience is built around consumption, but consumption filtered through the idea of "as if." The shops and products available frequently play important roles in the series. Being "able" to shop in the Wizarding World therefore also contributes to the illusion of "actually being there":

It literally is like you are in that shop. [. . .] It was just really cool, like I can't even believe I'm shopping in Honeydukes right now. (Hanna, 26, American)

Shopping is an immersive, imaginative act, one that connects Hanna to the story-world. She can visit the same candy shop that Harry and his friends visited throughout their time at Hogwarts, and purchase what she would if she were a witch. The stores are distinct, with different specialties that are frequently "diegetic extensions" as described by Mittell:[48] products from the series, or products that could be from the series (like shirts advertising in-universe sports teams), "released" into our world, which Williams discusses as a way to "encourage those within themed spaces to immerse themselves further in the constructed world."[49] This creates a more realistic, but also more touristic, sense of place:

Like you are just vacationing in Diagon Alley, and then occasionally popping over to Hogsmeade, like it really did [feel like this], because there is so much going on, there is so much to do, and it's just so cool, and it was really like vacationing . . . in . . . the Wizarding World. (Jen, 30, American)

When at the Wizarding World, visitors do what they would if they were visiting any other urban tourist destination: buy things, wander the streets, get something to eat or drink, perhaps see a performance or people-watch. These can all be done convincingly, with "local delicacies" like butterbeer and exclusive souvenirs, in a way that adheres to the narrative memories and details of the Harry Potter series.

The Wizarding World of Harry Potter became a massive success for Universal Studios. This popularity meant many other people in attendance, and sometimes long lines for rides. Crowds are often brought up generally as one of the reasons that theme parks aren't as fun as they pretend to be. It is interesting, however, that not every interviewee considered the crowds or lines a negative part of their experience:

> A lot of times we'll just go and sit on a stairwell or something and you just watched people going around, or witches and wizards going and doing their things. And just kind of pretending, that "Oh, maybe I am just hanging out in Diagon Alley and having an ice cream right now." It's just fun, it's total wish fulfillment. (Dan, 33, American)

Other people create the sense that the Wizarding World is a lively and living space, much as the locations are in the series. It becomes a place to "hang out" rather than a place to pass through. Many interviewees spent entire days in the Wizarding World section. The visitor takes on the role not only of a witch or wizard, but of an urban flâneur—strolling the streets and observing the other inhabitants of the city (who are also fellow characters in the story). To complete the sense of immersion, the space must be occupied.

The Social (and Fan) Space of Harry Potter

As testified by the interviews, another significance of others occupying the space derives from the sense that they were also there for Harry Potter:

> For the most part a lot of people that were there, they really enjoyed and loved the Harry Potter series. And so it was kind of like this, all time, 24/7 nerd convention inside there. And so you could sit there and freak out about a T-shirt because it was awesome and it was your house, and nobody would be like "oh my god that person is really weird." (Maggie, 21, American)

Maggie sees the space as a Harry Potter environment in which she can "geek out" about the series without feeling self-conscious. Rather than simply connecting to one's individual fandom, visiting became a way of performing it publicly and connecting to others who felt the same way. In the previous chapter, I talked about how fan conventions also foster this connection to other fans, becoming an important part of how the fans interviewed felt about Portmeirion. In these spaces, fans feel they can "be themselves" and embrace the "nerdy" interests that they feel they must hide elsewhere. Similarly, at the Wizarding World fans felt as if they could indulge their fandom in a way they couldn't in everyday life.

For some interviewees, being able to do this was also a bonding experience with the other fans in their lives. They came to the Wizarding World deliberately with others who were also fans of the series, and experiencing the space and the environment together enhanced their experience. As Hanna describes:

> [Going with friends is] probably the best experience you can have, because they are as into it as you are. I haven't gone with my family yet, but I hope to one day. Just because, like my older brothers are really into it too. I don't know if they'd be as "Oh my God" as we were, but it's just you feed off each other, that enthusiasm just stays constant [. . .] I would definitely recommend the first time you are going: go with people who are as into it as you are, because you'll have such a better experience. (Hanna, 26, American)

For Hanna, experiencing the Wizarding World with her childhood best friends, the ones that she had grown up sharing the series with, made it even more special. Their enthusiasm about being there and being in that environment after all the years they'd spent with the series enhanced hers, and she could recall their history with the series as they moved through and interacted with the space. Her fandom always had this social element, and visiting the Wizarding World with her friends made the sociality of their shared fandom part of this place as well. It was a time they could experience together.

The affective power and temporal element of being in the place with these important others also creates positive memories:

> It's one of those warm family memories that you remember over the time [the children] were goofing around and broke something, you

know, or the time they spilled the milk all over the floor or whatever. It's going to be one of those that carries forward. And it's going to carry forward for all of us. (Daphne, 44, American)

Daphne visited the Wizarding World with her husband, young son, and teenage daughter. The Harry Potter series had long been a point of bonding between her and her daughter, one that they were starting to share with her son as well. Sharing their fandom gave them something in common, an emotional connection to each other's inner lives that Daphne valued deeply as her daughter aged. Being together in the space of the Wizarding World and fully embracing its "as if" pretense together was another level of this shared connection, one that wasn't possible in their normal lives. Removing themselves from normality and into the enclosure of the park allowed them all to indulge their fandom without reservation and provided Daphne a memory to go back to when everyday life returned.

The assumed shared interest in the series can also create a sense of fellowship among visitors that don't know each other, at least among those exhibiting fannish appearance and behavior:

It was just really cool to have that kind of camaraderie throughout all spans of life. 'cause I mean, you could live across the world and we could both be in the same house, and we're like "Slytherin, what's up?" and you have that connection now with somebody that you had no idea who they were five seconds before. So it's really cool, it's like a giant family. (Circe, 29, American)

For Circe, a Florida resident who regularly visits the Wizarding World, the other visitors to the park are not anonymous strangers—they are other Harry Potter fans that identify with the series as she does, indicated by fan behavior like dressing up or showing knowledge of the series. The connection she has with the park's other inhabitants makes it feel welcoming and special. Being in the park means not only being surrounded by the narrative world, but being surrounded by this community. Contrary to Aden's view, what this suggests is that the Wizarding World is felt as a place of communitas as Turner[50] would describe it. Entering the Wizarding World area means that a fan becomes part of the "giant family" of fellow Harry Potter fans occupying the space, no matter their status outside the park, with the accompanying freedom to "geek out" and act like a fan in a way that transgresses society's normal proscriptions against such behavior.

There is also a sense among the fans interviewed that this was their space, one that was created with them in mind:

It's definitely a place where if you aren't a fan you can enjoy it, but the people who get the most out of it are the people who know what they're looking at. And I think that they really had that in mind. (Violet, 21, American)

While non-fans might visit, and even enjoy the park for the unique craft of the simulation, Violet suggests that those who truly appreciate it are the fans of Harry Potter. This is not so much an appropriation of space by fans, as is discussed in relation to other sites of film tourism or "fan pilgrimage,"[51] but a sense that they are in a place made for them. This works with the broader sense of communitas present among fans in the park. It is a space to perform fandom, made for fans, and therefore welcoming.

Conclusion and Future Directions

How do visitors interpret and experience simulated environments, and what leads them to embrace the Wizarding World of Harry Potter? These are the questions that I investigated in this chapter. As media-themed environments prosper, it is important to understand what visitors make of such spaces and why they continue to appeal, compared with the experience of filming locations in the previous chapters. In previous literature on fan tourism there has been regular discussion of why these places might be unappealing for fans, but this chapter suggests why they might be enjoyable as well. It also points to what kind of role these re-creations can play in contemporary fan culture, something that becomes important in chapter 5 as we look at another industry-created fan place. Through participatory observation and interviews, I have identified several key themes in the Wizarding World fan experience and have shown how it has become an important place for Harry Potter fandom.

First, I have shown how the theme park can be seen as a medium. Considering theme parks in this way moves beyond discussion of whether they are positive or negative for society and toward a consideration of what they are as a form. As a medium, the theme park has its own specificity—most notably separation from the "outer" world and physical interaction with a narrative. In theory, it should be sensorily authentic and provide a spatial and embodied connection to a narrative. This makes the theme park similar to ideas of immersion as found in media studies, where it is seen as a way to make a fictional environment feel "real," whether through technology or through the mental process of imagining the world as inhabitable.

It is this concept that the Wizarding World builds on, and that the fans responded to. For the fans interviewed here, the Wizarding World was interpreted as an adaptation of the Harry Potter series in its own right, one that uses the medium specificities of the theme park to present a new interpretation of the series. The fans have already mentally inhabited this narrative world through their ironic imagination, which is extended into the park. The approval of series author J. K. Rowling and the amount and quality of detail from the series found in the park gave it a sense of sensory authenticity. Through being there, visitors gained an embodied understanding of what it would be like if they could actually visit these locations. In doing so, they felt connected not only to the narrative itself, but to their fandom and the other fans inhabiting the space—they are all there for, and in, Harry Potter.

This study also shows the utility of the "ironic imagination" when discussing the theme park experience. While Saler speculated that "ironic self-reflexivity may be powerless against physical reflexivity,"[52] we can see from the example of the Wizarding World that even though the physicality makes the environment more "real," there are still limits. It is "as if" Diagon Alley is visited, and for all the emotionality of "being there," there is appreciation for, and awareness of, the work put into making the experience. It is this double consciousness that makes the simulated environment work. The structure and nature of the theme park—that it is a set-aside place of pretending—means that is interacted with and experienced as a fiction. Not every theme park will embrace this to the same extent as the Wizarding World, and not every visitor will accept the illusion as believable, but as the industry moves toward more immersive environments, it is important to understand the way that many visitors relate to and evaluate such spaces.

The immersive experience of the park and its status as an official transmedia expansion of the Harry Potter universe has proven to be a substantial draw for fans. It has come to function as a space of fandom, one that fans can come to in order to see and connect with other fans. This can be a connection with fans they already know, such as family or friends, or simply with the general fan community of the series. It is seen as a place in which fans can be themselves—they can geek out over a T-shirt or fully embrace the pretense of being a witch or wizard without feeling self-conscious. They are, after all, in a sealed-away place of pretending, deliberately designed to encourage a disconnect from everyday life. The role it plays for Harry Potter fans is not only as a place to physically experience the story-world, but as a place to act as a fan more fully than they feel they can outside of the park.

That the Wizarding World can play this role for at least some fans is an indication that it also exemplifies a new relationship between media producers and fandom. By encapsulating the fan in an official version of the Harry Potter world, it also absorbs practices that had previously been the domain of fans and brings them back under control. This is most clearly visible in the purchasable merchandise there, which draws upon traditions of cosplay and fan creation, but also in the Wizarding World's positioning within fan culture. At one point, Universal Studios worked with fan groups to host events like the Celebration of Harry Potter, a convention-like event with special appearances from actors and behind-the-scenes personnel, and the "Open at the Close" event as part of the LeakyCon convention of Harry Potter fans, but this seems to be no longer done. Rather, the park is situated as the site of fandom based on the experience of the park itself, the place that fans want to go to in order to feel more deeply involved with their fandom and pay tribute to it. While fan conventions for the series still exist elsewhere, the existence of the Wizarding World means that the conventions' power of being "the place" of fandom must be shared with Universal Studios and, to some extent, with other controlled spaces like the Warner Bros. studio tour and Platform 9¾.

The fans involved in this study (and it must be noted that these fans were already inclined toward appreciating the Wizarding World) suggest how this relationship works. They were pleased that Universal had put the resources into building the park in a way that meant they could have the desired embodied—and official—connection to the story-world. That Universal had done so was a sign that their fandom was valuable and valued. They were happy to pay for the experience, which they found meaningful, and were reluctant to critique it too heavily. If Jenkins could see fans in 1992 as rebellious poachers, working at a counterpoint to the media industry,[53] fans in the contemporary media environment might better be seen as the game. They are carefully cultivated, valued, and used for resources. This is not to say that "poaching" as Jenkins described it has disappeared, as it was always focused on the mental processes and imaginative work of fans in interpreting industry products,[54] but that the relationship has changed, at least for highly successful media franchises like Harry Potter. This has potential repercussions for contemporary fandom, as I will look at in the next chapter through a study of how fandom is imagined at events for the television show *Friends*.

What is also worth taking into future consideration is the role of the park in the expanding Harry Potter franchise. At the time of fieldwork, there had been little new content in the story-world, following the conclusion of the film adaptations. Since then, however, there has been a clear attempt to make the story-world into a true franchise, akin to *Star Wars* or the Marvel Cinematic Universe. A sequel stage play, a multiplayer mobile game, and a blockbuster prequel series, set in 1920s New York City, have all been released, but not without controversy in the fan base and the general audience despite Rowling's involvement in all of them. However, the park continues to be a draw.[55] As the story-world becomes a franchise, it is likely that place will be one of its most important—and lucrative—nodes. This suggests that in studying contemporary franchises and their fans, the theme park should not be thought of as an afterthought, but as an increasingly integral way of interacting with a story-world. In the next chapter, I investigate some potential implications of this development for fandom more generally.

Friends and Corporate Places of Fandom

It is, possibly, the most popular television show of the 1990s, and one with a surprisingly long afterlife. Fans spanning generations continue to wear its merchandise and make reference to its catchphrases. It has been a site of a major battle between streaming services, with Netflix spending an estimated $80 million to $100 million to keep the show on its roster in 2019,[1] only for it to leave soon after to anchor HBO Max. I am, of course, referring to *Friends*, the American sitcom about six young people living in New York as they figured out their place in the world and their relationships to each other. A massive success for its ten seasons on air, both in the United States and abroad, it has found a seemingly eternal life as a staple of reruns and streaming services, and has become remarkably popular among those who weren't born when it first aired.[2] Twenty-five years after it premiered, *Friends* seems stronger than ever.

Despite its popularity, *Friends* was not always thought of as a show with a fandom. It had fans, certainly, but not the sort of structured, "cult" fandom that is demonstrated in the other chapters of this book. Rather, *Friends* was held up as the opposite of these texts and their fans—it was mainstream, accessible, and everywhere, without the exclusivity and transgressive potential that cult fans see in their objects of fandom.[3] It did not have a presence in the fan conventions and message boards that made up what is commonly known as "media fandom." There was, after all, little potential stigma in watching one of the most popular shows on television, and little subcultural cache to be gained by supporting it. It operated in a different sphere of culture.

However, attitudes about fandom have shifted, and the borders between mainstream and fannish activities are now blurry. Being a fan, at least of certain texts and objects, carries less stigma, and at the same time, some practices of media fandom have spread out into other environments. There is recognition in the media industry that fandom can be of benefit to their products,[4] through promotion and financial support, a trend that has accelerated as media audiences become more fragmented and there is an increased desire to hold on to the audiences that a show does manage to get. Having fans is now a way to continually generate attention to a media text and, through that attention, income and prominence.

This doesn't mean that every aspect of fandom is desirable. Media fandom and its practices developed at a time when the industry took little notice of what fans were doing. This neglect led to a situation where fans developed a wide variety of creative practices that made transformative use of their favorite texts, from fan fiction writing to criticism to cosplay, often at odds with what the industry wanted.[5] The structures of fandom itself were seen as an alternative space to challenge social norms, a mindset that often developed from such creative endeavors, existing outside of a commercial field of value and as an alternative to it.[6] Now the industry is paying attention, and fandom is (relatively) mainstream. Traditional "grassroots" fandom and corporate engagement strategies now operate in the same space. It is therefore worth investigating what aspects of fandom are encouraged by these engagement strategies and what impact this might have on how we think about fandom.

As previous chapters have shown, place is an important part of contemporary fan practice, a way to physicalize the connections and relationships that fandom makes, making the alternative world of fandom real. Connecting with places means connecting with fandom, and can be a powerful, and highly sought after, emotional experience. The practice has risen in prominence and visibility, something that has not gone unnoticed by a media industry increasingly looking to create "experiences" around texts as a way of developing engagement and attachment to a brand through interaction.[7] After all, as the previous chapter suggests, fan places can be created and controlled by the media industry without too much resistance from fans. Tourism and place are therefore one of the central locations where the contemporary struggle for fandom is fought.

This creation of place is the focus of this chapter. The continuing popularity of *Friends* has led to the creation of several spaces as a way to capitalize on the interest in this direction. The first, FriendsFest, is a touring "festival" organized by cable channel Comedy Central UK that originated in London six years ago as a pop-up and has since spread across the country and into Spain, Poland, Germany, and Russia. The second, the *Friends* pop-up in New York City, ran for a month in the fall of 2019 as part of the twenty-fifth anniversary of the show's first airing. Both places feature re-creations of sets, photo opportunities, and other activities, run by professional experience marketing companies under the supervision of Warner Bros. and/or Comedy Central UK. They are fan places, designed with fans in mind—but deliberately created ones, rather than places appropriated by fans. What kind of fandom is encouraged by such corporate places of fandom? As fandom is increasingly commoditized and made mainstream, how can place be used to shape fandom? These are the questions that I look at here.

While the previous chapters focused on interviews with fans, this chapter will instead focus on the marketing and promotion of place to fans. *Friends* makes a particularly interesting case subject not only because of the success of the events (both the pop-up and FriendsFest have been sellouts), but also because of the nature of *Friends* fandom. Without the structures of cult fandom, as can be found in the previous three chapters, the organizers of FriendsFest and the *Friends* pop-up are, in some ways, building the organized fandom of the show. But how do they see this fandom? What does it suggest about what kind of fandom is promoted and accepted by the media industry? Through this investigation, I show how place becomes part of "fanagement" strategies[8] in the age of the "experience economy," a way of fostering engagement with a text as a brand through unique "experiences" presented as an opportunity for devoted fans to interact with the text in a way that is "more than" just watching. Ultimately, though, these experiences reinforce particular producer-friendly readings of the text-as-brand, under control of the producers, rather than more appropriative uses of the text. The marketing and promotion around the *Friends* events provides an ideal starting point to investigate these developments, as it is through these kinds of paratexts that the strategies to attract the "right" kind of audience are played out and ideas of appropriateness are reinforced.

Friends, Fandom, and Contemporary Marketing

Much of the earliest research in fan studies begins from a stigmatized place. Jenson's classic work "Fandom as Pathology: The Consequences of Characterization" explains this idea in depth, laying out the descriptions common at the time in popular and academic psychology and social science of fans as obsessive, deviant, and otherwise pathological, "irrational, out of control, and prey to a number of external forces."[9] They are over-consumers of trash culture, easy led by the media industry, and lacking in social skill, preferring their collections and daydreams to participating in the real world. They were annoyances to media producers, who didn't want to be associated with this kind of behavior. "Fans" could also be compared unfavorably with the more acceptable "aficionados and hobbyists," who, while displaying emotions and behaviors similar to those of fans, were seen as "normal," even admirable, as their interests are of higher social status or show some level of acceptable sociality, a point echoed in Pearson's "Bachies, Bardies, Trekkies, and Sherlockians"[10] and McKee's somewhat tongue-in-cheek "The Fans of Cultural Theory"[11] published a decade and a half later. Fans, at that time of writing, were a "them" to most, compared with the "us" who had proper interests and boundaries. Fandom was a strange behavior, and identifying as a fan was something that people only did if they were willing to accept the idea that they were not "normal people."

We see a shift beginning in the 2010s. In academic literature, the rise of fan studies presented fans in a different light. Rather than deluded over-consumers, fans were instead "poachers" who took elements of popular culture and refashioned them in their own image, creating alternatives to the mainstream media industry that highlighted diversity and viewer creativity. They were discerning and supported quality texts, often quite obscure ones, rather than mainstream successes. They were also social and community-oriented, with the structures of organized fandom—conventions, message boards, and so forth—providing an alternative social space for often marginalized people. Just as importantly, fandom did this outside of a commercial, commodity-driven market. While fans used corporate popular culture, their own works and activities stood outside of it, operating on a "gift economy" model[12] that encouraged connections with other fans as much as with the original text. Fandom in this framework is an alternative way of interacting with media, one that encouraged active, creative interpretation of the source text, social connections with other fans, and a noncommercial ethos. You didn't have to pay to be in fandom.

The more positive interpretation of fans in this period was not limited to academia. Geraghty discusses how media about fans in the 2000s, while drawing on the classic stereotypes, should nevertheless be "seen as celebratory of what it means to be a fan in a networked and socially connected community."[13] While the fans depicted might be socially awkward geeks, Geraghty argues that they are also, often, ultimately heroes of their stories, with their fan identities and collections bringing them happiness and success by the final scenes. Mockery of fans is affectionate and often grounded in the creator's being part of the fandom, able to share in the in-jokes and revel in embracing, as well as overcoming, the stigma of fandom. This is one indication of how the concept of "fan" is an increasingly visible and generally positive one in contemporary society, at least for certain conceptions of what a "fan" is.[14] While maybe a little bit out of the ordinary, fans are now seen as passionate and interesting people, able to discuss popular culture in depth and maybe even to transform that ability into a satisfying creative career. Love of and deep investment in mass culture moved from a pathology to an interesting personality trait.

The decrease in stigma also means that fandom is a lower-stakes position than it was in previous eras, not necessarily demanding the kind of subcultural identity as it might once have. In the words of Booth, "as fandom becomes more diffuse, more media viewers are considering themselves 'fans.'"[15] As I discuss in chapter 1, there is a recognition that fandom is inherently based on emotional involvement with particular texts, which, rather than an extreme position, is one that is very familiar to most. With the internet and social media commonplace, "participating," in terms of engaging with extra content about a text, or discussing it online, has also shifted from a niche behavior to an expected one. It's not particularly difficult to create a social media profile and discuss an episode with other fans, or to seek out news about a television show, or to pass along a BuzzFeed quiz that tells you what *Friends* character you are. The infrastructure of social media was largely built on what fans were already doing, both in person and online,[16] meaning that some—although certainly not all—fan practices are now normal, just part of how we interact with popular culture online. To some extent, this means that considering oneself a fan is as well.

The term "fan" thus becomes less of a specific subcultural identity, but a "catchall term for interactive consumers,"[17] in a way that, as Stanfill argues, suits the media industry's needs in a complex environment. Fans are now a "constituency that media companies both recognize and actively seek to incorporate, encourage, monetize, and manage."[18] In an attention economy, having fans as the media industry might describe them—engaged and active consumers—is a cornerstone of survival and success, and what makes it possible for a show to transition to a more lucrative brand. Fans are the ones who will not only buy the merchandise, but support and promote the text to others, drawing in more audiences by their freely given labor.[19] Instead of nuisances, fans can be useful and productive for the industry, and are seen as something to cultivate and manage in particular ways. They are what make a text matter over a longer term—as we see clearly in chapter 3. Fandom is also no longer limited to media texts. Encouraging fannish identification is part of the marketing strategy for nearly every kind of product and brand, from hotels to sportswear. As Linden and Linden emphasize, contemporary marketing suggests that "being normal equals being a fan, and proclaiming it as often and as loudly as possible is 'good' social media behaviour."[20] People should be fans of a brand and identify themselves as such to others in a way that makes them advocates for it. This is what makes an "ideal consumer."[21]

The current state of *Friends* fandom is a particularly interesting illustration of this shift. The show premiered in 1994, a very different media environment than today. As a sitcom about six young people living in New York City, it was outside of the kind of shows that the structures of fandom at that time tended to support, not only in terms of content, but in that fandom saw itself as irreparably different from the mainstream.[22] *Friends*, from its beginning, was distinctly mainstream. It was a ratings success from the outset, quickly cementing its place in NBC's "Must See TV" Thursday night lineup and anchoring it for the decade that it aired. Its success and approachability were its defining factors—the series was easy to get into and easy to like, with attractive characters, witty dialogue, and low stakes, a sort of "comfort food" of television,[23] validated through its high level of worldwide popularity. Compared with its contemporaries, such as *Buffy the Vampire Slayer*, *The Sopranos*, and especially fellow New York–set popular NBC sitcom *Seinfeld*, with which it bears some textual similarities, it was not considered part of the "quality" TV discourses that arose at this period, and rarely received critical accolades.[24] It was, in many ways, emblematic of "normal" television, rather than the quality or cult discourses that fans saw themselves as participating in.

However, as Todd[25] and Kutulas[26] discuss, *Friends* as a show was structured in a way to encourage "fannish" engagement. While not quite serialized in its storytelling, the show saw the characters develop over time, growing and changing in their relationship to the world and, most importantly, to each other, with four of the six leads ending up together by the close of the show. It was a show that rewarded regular viewing, with recurring catchphrases and callbacks to previous episodes, and one that encouraged intimacy and identification with its characters and settings. Avid viewers of the show wanted to be one of the *Friends* characters, hanging out in the coffee shop and having hilarious experiences in the big city, and found themselves identifying with their personalities and issues. While not being part of the cult media infrastructure, it nonetheless behaved much like a cult show textually, and its fans loved it in the same way.

In its first airing, this was enough—the program was a massive success for NBC and Warner Bros., with high viewership throughout its ten seasons and strong international performances as well. However, rather than fading away from public memory in the years since it wrapped, *Friends* has found a new life, and a new fan base, in the streaming era. The textual qualities that fostered its original success—the intimacy with, and development of, its characters, the causal "hangout" setting that made it easy to watch multiple episodes in a row, its combination of accessibility and aspiration— enabled it to continue to attract audiences[27] used to a different kind of television delivery. It has been a considerable success as a streaming property and has a dedicated audience among younger viewers who missed it the first time around.[28] It is clear to Warner Bros. that *Friends* has continuing value, and that there is worth in transforming it from simply a television show to a broader "brand."[29]

Johnson discusses how branding in television works to foster a stronger sense of loyalty to a program or network, giving an identity beyond the screen and encouraging further consumption over a longer period, particularly through merchandise,[30] something that these new, younger viewers have come to expect and that older viewers are delighted to now have (and tied closely to how fans are seen to behave). *Friends* as a brand has not faded post-finale, and, indeed, has grown even stronger, with merchandise lines debuting recently in Primark, Pottery Barn, and Urban Outfitters, new tie-in books coming out, and considerable attention paid to its recent twenty-fifth anniversary. The branding strategies encourage continued engagement with *Friends* as well as nostalgic recollections of what the text "meant" to fans, confirming particular story lines or images as important, and give fans a reason to consume the text once again and deepen their attachment to it.[31] Geraghty[32] conceptualizes nostalgia as a "tactic" of fandom, a way to keep a beloved text or object as part of one's life throughout different stages, triggered not by unmet longing but by exposure to the actual object in a way that encourages remembrance. In the contemporary media environment, older, "nostalgic" media is everywhere, as are interfaces to it in the form of websites, reruns, and now streaming media. Texts are always waiting to be remembered, revisited, and reincorporated into the life of the fan. *Friends* might be over, but it is never actually gone, and there is always a reason to return to it. This can work just as well for fans who watched *Friends* when it first aired and those who came to it later. The desire to return, remember favorite moments, and perhaps even find new things in *Friends* is shared throughout the different generations of *Friends* fans and the paratexts that cater to them.

As with other fannish pursuits, this can be encouraged in the grassroots manner Geraghty discussed, such as through fan-made websites and uploaded videos, or by media producers through strategies like the media anniversary. Indeed, the *Friends* pop-up in fall 2019 and the expansion of FriendsFest to Germany, Poland, and Russia were justified through anniversary logic. Hills discusses such anniversaries as part of branding strategy, but one that "position[s] it as an issue of (fan) cultural capital rather than as a purely commercial venture,"[33] as it is fans who give this brand anniversary meaning, even if it is instigated for commercial purposes by the producers. He stresses the need to take a "both/and" approach to these events—their commerciality does not mean they aren't meaningful to those who participate in them, similar to how I described the Wizarding World of Harry Potter in the previous chapter. It is the mix of corporate effort and fan-produced meaning that gives anniversaries their appeal as events. They encourage a particular kind of remembrance, but it is the fans' acceptance of this framing that gives them validity.

Additionally, events like FriendsFest and the *Friends* pop-up also draw on the concept of the "experience economy,"[34] a marketing discourse that argues that contemporary consumers don't just look for products, but seek experiences—memorable activities that create meaning. Compared with products, experiences are intangible, based in sensation, and can only be accessed in the moment, but are increasingly sought after as consumers get used to having similar products always available. Experience sets businesses apart from each other and creates deeper engagement with and loyalty to a brand, giving consumers something memorable, fun, and emotionally affecting that is then connected to it. More importantly, people will pay more for an experience than simply for a product.

As I discuss in chapter 1, this is the inherent premise of tourism, and particularly film and fan tourism. Tourism is a textbook example of an experience,[35] as it is intangible, in the moment, dealing with the senses, and, ideally, meaningful. People are willing to pay for the experience of being somewhere. This extends quite easily to film tourism, in its creation of an experience with a text that goes "beyond" watching. While originally non-commercialized, the experience can also be commoditized and used to add value to a brand,[36] as I showed in the previous chapter. Place markets an "experience" to contemporary consumers, one that is potentially quite lucrative in terms of direct consumption while there, but also in continuing to build the brand and further incorporate fandom into industry practice rather than letting it exist outside of the commodity system.

Booth, in the context of Cardiff's Doctor Who Experience, thus refers to industry-created places as a "media parody," a co-option by the industry of fannish affection and play.[37] In "parodying" what fans were already doing, media producers "both enable and constrain fan audiences,"[38] giving them new experiences but diverting them toward commercial, rather than community, ends. However, as the experience at the Wizarding World of Harry Potter shows, that doesn't mean they can't be legitimate places of fandom for visitors. The Wizarding World matters to fans, and the experience is an overall positive, emotionally resonant one that connects the fan to their fandom and the positive role it has played in their lives.

The Doctor Who Experience and the Wizarding World of Harry Potter are interesting examples of how the media industry creates experiences for different eras of fans. *Doctor Who* has a long history as one of the archetypal "media" fandoms, meaning it has a relatively smaller but devoted audience that creates a community based around their interest, with conventions, creative works, and a sense of shared identity. The Harry Potter series combined massive mainstream success with a large fandom community in this tradition, which produced a great deal of creative works and a sense of community among its most devoted fans.[39] Comparatively, *Friends* fans are only recently seen as part of a fandom. Promoting its places is as much of a construction of fandom through use of place as it is a parody of it. But what kind of fandom is it? This is where we now turn.

Methods

This chapter has a different focus than the others in this book, and therefore it requires a different methodology. Rather than looking at the film tourists themselves, I instead look at how these places are marketed and promoted to them through the media. As Bennett argues, media depiction of fandom is a key site for the way in which fans are constructed and understood by the general public and by fans themselves.[40] It is through media framing that we can look at the ways in which fans are understood in the world at large, and how the media industry specifically might see them. This is particularly true for the articles I analyze here, as they function largely as promotion for FriendsFest and the *Friends* pop-up and are designed to attract a particular audience. This sort of "hype" prepares an audience for what to expect with a given text and provides guidelines for how to interpret it.[41] Through looking at these articles, we can see how the entertainment news industry and, often, the places' creators think of the fan audience, in their inclusion of quotes and appearances from them. I have focused on news media, rather than social media, in order to better explore the professional framing of these places. This chapter is therefore designed as a reflection of the developments that I explore in the previous chapters, showing one way in which media producers and the broader entertainment news industry are reacting to the rise of film tourism.

I collected thirty articles about the *Friends* pop-up and seventy-six articles about FriendsFest from its origins to its most recent edition, through a combination of LexisNexis and Google searches. This amount allowed me to get a general sense of the coverage, while still allowing me to analyze the texts closely. I focused on written professional publications, rather than blogs or other "fan"-produced writing (to the best of my ability to discern this). Both events were covered by a mixture of local, national, and in some cases international coverage (particularly toward the pop-up), in the entertainment, celebrity, and event calendar sections of professional publications. Additionally, I consulted sources connected directly to the events, such as the websites and FAQs for each event, the Comedy Central UK coverage of FriendsFest, and websites of some of the companies involved with putting together the events. The articles on the *Friends* pop-up span from its announcement in July 2019 through its opening and presence in New York City in October 2019, while the FriendsFest coverage begins in 2015 with the first London version and continues to the late summer of 2019. While there have been versions of FriendsFest in Moscow, Warsaw, Düsseldorf, Barcelona, and Madrid, I limited my analysis to English-language coverage and the English versions of the events, as that is the language and context I understand the best.

I used a strategy of rhetorical analysis to understand not only what is being said, but how it is being said, and perhaps just as importantly what is not said.[42] These pieces are designed in order to promote these sites and *Friends* fandom, and therefore depiction of fans and what they do at these places has a rhetorical power. It is not just description, but also a particular construction of *Friends* fans and fandom, an argument that this is how *Friends* fans should feel toward and behave at these sites. Many of the articles, particularly those in local publications in the areas that FriendsFest visited, are remarkably similar, showing the influence of PR professionals and press releases in the coverage. They often feature quotes from official sources, including executives behind the event and actors brought in for their promotion. Therefore, they are part of the promotional and marketing process for *Friends* places, showing not only how the media industry sees fans of *Friends*, but how it constructs a fandom out of what was once just very popular.

The Presentation of Fandom in *Friends* Places

Normalizing Fandom in Friends *Places*

FriendsFest is for fans. This is clear when browsing articles about it, from the first pop-up in 2015 to the upcoming (at the time of writing) Friends-Festive extra holiday event in London. It is "every *Friends* fan dream come true,"[43] the "ultimate experience for the *Friends* ultra-fan,"[44] or simply "the one where fans wander around Monica's apartment."[45] This is a view shared by the creators of the event, who say that "fans will be able to fully immerse themselves in the world of *Friends*" in the very first words of the FAQ on the official website. Similar terminology is used when promoting the New York City *Friends* pop-up, which is for "superfans of *Friends*,"[46] "targeted to old and new fans alike,"[47] and asks "could *Friends* fans be any luckier?"[48] The word "fan" is used throughout, both as an attention-grabber and as a banal description of whom the places are for, placed in the headlines and body of the lighthearted profiles and promotional stories about the events.

From the outset, it is clear that FriendsFest and the *Friends* pop-up are designed for fans and meant to appeal to those who consider themselves *Friends* fans. They are not addressing a casual viewer or looking to attract new fans to the program. The goal is to appeal to and attract existing fans of the show, reminding them of their fandom and renewing their interest in it. In descriptions of the offerings at these places, there is considerable attention paid to what fans are expected to want to see—noting the props brought in from the original filming, making reference to particular episodes when discussing what can be seen at the events, and advising attendees to look for particular Easter eggs in the re-created sets (such as a copy of *The Shining* in Joey's freezer). Mentioning these smaller details signals that, similarly to the Wizarding World of Harry Potter, these spaces have been made "for" fans, as it is fans who will seek out these details, and evaluate the places based on whether they are present or not.

What is interesting here is not simply that these are fan events, but that this is a positive thing. There is no trace of fan stigma, even the mocking-but-affectionate kind that Geraghty sees in his exploration of fan-focused fictional media. While a small number of texts are critical of the event themselves, with the *New York Post*[49] and Slate[50] decrying its "fakeness," this is not done at the expense of the fans who go or want to go. Rather, fandom of *Friends* is depicted as a completely understandable, positive thing. At the press preview of the *Friends* pop-up, for example, actress Maggie Wheeler, who had a recurring role on the show as Chandler's girlfriend Janice, stated that "*Friends* has kept people company at such a deep level, through good times and bad times, through heartbreak and sickness, and through all kinds of life chapters. People have so many points at which they return to this show, and so it's no surprise that everybody wants to come see this in real life."[51] Fandom of the show across the fan's life span, extended to the pop-up, is treated sympathetically and seen as valuable, something that helps fans, much in line with the concept of fandom I discuss in chapter 1. Expression of this view by someone who was part of the show, as part of the promotion around this official event, shows the level to which a positive depiction of fandom is part of the way in which it was conceived and promoted.

This extends to media depictions of *Friends* fans, even ones that might be thought of, in other cases, as more "extreme." For example, a MetroUK article about a couple who got engaged at FriendsFest in London in 2016, where they "basically recreated Monica and Chandler's proposal"[52] at the re-created apartment set, is positive toward the couple and their experience. While they are described as an "obsessed couple," and quoted as saying "Radha can't go to sleep without watching the show," the engagement is treated as an enviable romantic occasion rather than an occasion to gawk. "Talk about boyfriend points!" says the article's author when describing the future groom and his proposal in detail, particularly admiring the effort that he took to ensure FriendsFest's cooperation and involvement, including removing others from the set for the proposal itself and having photographers and champagne on hand. It is a quirky, but ultimately approved, experience.

In comparison to the "nerd-focused" texts profiled by Geraghty, or even in Booth's description of the Doctor Who Experience, there is little playful fannish self-mockery in the articles. While *Friends* fans might be a little bit "geeky" about *Friends*, it is a mainstream show. Being a fan of it is not seen to have the stigma of "traditional" fan culture, or the pride in overcoming that stigma, and therefore what is left when discussing the fans is enthusiasm, which is assumed to be shared by others reading these articles. This comes through quite clearly within the articles, with *Friends* described as featuring "our six favourite TV friends"[53] and remaining "one of our favourite shows of all time. STILL."[54] There is an assumption that love of *Friends* is a shared trait in contemporary society. The editorial tone toward the events is therefore also enthusiastic. MetroUK says that "we actually cannot contain our excitement" that FriendsFest was returning in 2017,[55] Delish describes the pop-up as "every bit as magical as we hoped it would be,"[56] and BuzzFeed posted an excited, picture-filled list of three of its staffers attending the New York City pop-up and participating in all of its activities.[57] Being excited for, and enthusiastically participating in, the *Friends* places is not something for "them," it is something for "us."

This normalcy is also reflected, at least in the UK publications, by showing celebrities at FriendsFest. Gala openings and/or closings in London and Manchester attract celebrities and press coverage, and the celebrities' fandom of *Friends* and participation in FriendsFest activities are regular topics of coverage. The kinds of celebrities that show up in the coverage—mostly reality TV and soap stars—are ones that rely on a certain amount of not only visibility, but relatability and accessibility, suggesting that showing up to these events at the very least causes no harm to their reputations. While there, they are shown posing on the sets and furniture "as" fans and participating in the various scene reenactments set up as part of the event. There is little to distinguish these photographs from those posted by noncelebrity fans, other than their placement on the "Celebrity" section of the news sites they are found on. Some go further and display their fandom quite visibly, wearing *Friends* clothing or describing themselves as "fangirls."[58] Displaying fandom so publicly signals that being a *Friends* fan is positive and relatable.

In general, this is what the overall tone of the press coverage suggests. It is stressed that the imagined audience of these places consists of fans, rather than those with a more casual interest in the series. Alongside this positioning, there is a normality to the way in which *Friends* fandom is depicted in coverage of FriendsFest and the *Friends* pop-up, compared with the traditional othering of fans and fandom. *Friends* fans are treated sympathetically and in casual, inclusive language that often suggests that both the author of the article and the reader are part of it. In discussing and promoting FriendsFest and the *Friends* pop-up, "everyone" is a fan—and therefore everyone wants to be there. It is perfectly normal.

Performing Friends *Fandom in Place*

The idea of fans as "active" is usually seen as the separating point between fans and "normal" viewers. Fans do "more" than watch—they make things, they go in-depth into their favorites, and they perform activities relating to their fandom beyond simply watching on a screen. This is what makes them fans. This understanding is also key to the conception of fans at FriendsFest and the *Friends* pop-up.

Both FriendsFest and the *Friends* pop-up are built around the re-creation of the key places of the sitcom: the apartments of the *Friends* characters themselves and the coffee shop where they meet. As a traditional multi-camera sitcom, the show was built around particular "rooms,"[59] which are re-created to differing degrees at the different events. The larger and more established FriendsFest began in 2015 with a re-creation of the apartments of Joey and Chandler and Monica and the Central Perk coffee shop, and now boasts walkthroughs of all the main characters' apartments, while the *Friends* pop-up features Joey and Chandler's living room, Central Perk, and more "authentic" props. As Knox and Schwind[60] and Thompson[61] discuss, the places of *Friends*, especially the domestic interior of Monica's apartment, are key to the storytelling and feel of the show. Compared with cinema or even contemporary single-camera sitcoms, which move about a space and might even film on location, *Friends* largely takes place in these staged domestic and semi-domestic interiors, with a similar view of them in each episode. They are therefore extremely familiar to fans. It is these places that "are" *Friends*, and that fans are assumed to be attached to.

Fans are thus invited to "explore" and inhabit them, an activity that is presented as a fulfillment of an immersion fantasy, particularly for Friends-Fest's larger set tour. The *Friends* pop-up, with fewer set spaces, plays on the volume and authenticity of its gathered props, with series actor James Michael Tyler (the barista Gunther) quoted in the Associated Press article on the event as saying "this particular pop-up is far more elaborate than the ones that we've had in the past. This is literally like a 'Friends' museum. It's amazing how many props and little things that just bring back memories to me."[62] This connects the *Friends* pop-up to the rhetoric around popular culture exhibitions, drawing on their ideas of authenticity and value to present a direct physical connection to *Friends*. Fans can admire these relics of filming and use them to further understand what the show was "really like" while also recalling their own history with it. Fans are constructed as desiring the ability to see more of the places and objects, in more detail.

"Exploration" comes up frequently in discussing both the pop-up and FriendsFest, aligning *Friends* with the larger rhetoric around transmedia storytelling. Originally developed by Jenkins to describe the kind of contemporary media storytelling that tells a continuing narrative across multiple forms of media,[63] transmedia storytelling is largely associated with the sort of "cult" texts that, as discussed earlier in this chapter, *Friends* has generally been constructed against. However, if we look at what Stanfill calls "transmedia dimensionality,"[64] we can see how FriendsFest and the pop-up encourage a certain sense of transmediality with *Friends* as well. Transmedia dimensionality is designed to provide more depth and breadth to a franchise, with the idea that there is "more" to explore in a given world—and that fans, especially, want this. As Stanfill elaborates, those interested in going "beyond the screen" are particularly courted by marketing companies (such as Superfly and Luna Cinema, the companies that put together the *Friends* pop-up and FriendsFest, respectively), and are "constructed as the ideal"[65] of contemporary fans, as it is these fans who will spend more time and resources on a story-world. They want to "explore" and "participate," a mode of fandom that requires performance and action.

This is demonstrated by how fans behave at fan places. As discussed in chapters 2 and 3, reenactments are a standard part of film tourism. They are seen as a way of connecting the tourist with the place, giving them the physical connection to the text that is often sought with these visits, and are often the most spectacular aspects of a visit to a filming location, as highlighted by their prominence at PortmeiriCon and TitanCon. They are part of what marks out a place as a fan one, in that they are acceptable places to perform fandom without it feeling strange, and by performing, the fans appropriate these places. Fan-run events stage reenactments, sometimes (as in PortmeiriCon) even providing fans with materials to participate more fully, and individual fans strike poses at particular locations to further establish that they are there. The action connects the fan to what happened there, and acknowledges that it is important to both the individual fan and other fans they might know.

Perhaps unsurprisingly, then, reenactments feature heavily in the descriptions of FriendsFest and the *Friends* pop-up. While the various editions feature a range of different reenactment spots, several are treated with particular reverence and appear in multiple editions. The "Pivot" scene, where the *Friends* characters attempt to get a couch up the stairs to Ross's apartment, voted "the nation's favourite Friends moment by Comedy Central UK viewers,"[66] appears at both FriendsFest and the pop-up with the event providing a large couch, re-created stairs, and directions to perform the scene accurately and stand in the right place for photos. The title sequence, again with props provided, is highlighted. The "Thanksgiving floating heads moment" and "the iconic three wedding dress shot,"[67] playing foosball at Chandler and Joey's,[68] and sitting on the furniture in the apartments are also prominently featured across editions. And of course, Central Perk and its orange couch are also mentioned as a place to sit and pose with a (purchased) coffee. They are all presented as "fan favorite" scenes that fans "have the opportunity" to participate in, adding to the idea that these are the "ultimate" places for fans, the correct place to perform their fandom.

They are also often described as "photo opportunities" as much as reenactments, designed and discussed as not just for the fan to experience in the moment, but to record (and hopefully post for others to see). The pop-up, for example, is described as a "tailor-made social media experience"[69] and highlighted as "Instagram-worthy!"[70] As discussed in chapter 1, the dialectic between photography and being there is a crucial driving factor of tourism and especially film tourism. Social media has only accelerated this, by providing more instant gratification and recognition of the experience to others.[71] It is important not only to do something, but to be seen doing it, and seen in a way that is recognizable for others—after all, isn't everyone a *Friends* fan?

The similarity between the events, in terms of re-creations and activities, points to a consensus as to what fans "want" to see and do. Both events feature set re-creations to explore, promising immersion and detail, similarly to the Wizarding World of Harry Potter in the previous chapter (a more classical transmedia franchise). The choice of specific reenactments that are encouraged and provided for across events reinforces that they are iconic and important scenes. There are actions designed for each spot, with particular props provided—twirl umbrellas here, pick up the sofa here, pose here—and sometimes even with directions provided. While it might be possible for other actions to take place in the space—we have to remember the appropriations that fans make in these kinds of places, as I talked about in the last chapter—certain actions are clearly set up as the correct ones to do, with a correct way to do them. They are reproduced in the photographs, both by reporters and by celebrities attending the events, highlighted in the texts, and tied back into the show as key moments, which are then highlighted in merchandise lines. Fans are encouraged to connect their nostalgia and personal memories of *Friends* to these particular "favorite" scenes, which become representative of the brand. Doing these things in this space is designed to be representative of one's fandom. Participating (and taking pictures) is how fans show that they are, indeed, fans of *Friends* who were "lucky" enough to be there. Performing proves their fandom by showing that they can do these actions in their correct place.

Stanfill notes that the interactivity provided by official extensions tends to take the form of activities that focus on "point and click and be entertained and as a choice within precoded options,"[72] such as polls, quizzes, and games. While these require more activity from fans than watching a show or film does, and provide for the fans who want to do something, they are still ultimately under industry control and depend on industry resources. They foster further consumption of the text, with more time and effort put into doing so, rather than appropriation of the text. This is a point echoed by Linden and Linden, who stress that "the ideal type of creativity in consumer society builds on existing frameworks," with the world being "like a computer game with pre-coded options to choose from. They may seem vast, but they are limited."[73] Fans are encouraged to do things, but not change things.

We can see this dynamic at play in the *Friends* places. The reenactments are much like the polls, quizzes, and games of digital extensions—they are imagined as happening in set ways, with set actions, and without space for the fan to appropriate or transform the story. This is not to say that individual fans might not do something different, but the ideal way for fans to interact with the places is the way that Superfly or Luna Cinema intended, and these actions are highlighted in the promotion for the events. Fans at FriendsFest or the *Friends* pop-up can interact, but in specific ways, and usually under supervision (for example, at FriendsFest, the set tour is guided). Comparatively, the Wizarding World of Harry Potter, while being similar to FriendsFest or the *Friends* pop-up in terms of what fans do, is seen as a place for fans to wander around in and come up with their own reasons to be part of the story-world in the park. The *Friends* places have much more structure. The ways of interacting with them reinforce the brand identity of *Friends* as a franchise by focusing on specific iconic moments, and by limiting the fans' ability to develop new stories while there.

Stanfill notes that "new norms absolutely include things fans do, but selectively."[74] This is evident in looking at the ideal fan performance in place. The reenactments, in many ways, resemble the ones found in the previous chapters. After all, fans also script reenactments, as I show clearly in chapter 3, where the main reenactments of PortmeiriCon are ones that have been performed for years, with props provided for new fans and set actions. Many *Game of Thrones* fans might also want reenactments as easy to participate in as ones found at FriendsFest, with such clear instructions and obvious iconicity for those back home. Fans want to explore the places they visit for fandom and want to see both familiar and new sides of it. The point of created events and places is not so that "fans cannot or will not take, but so they do not have to bother."[75] Fans don't have to create their own version of *Friends* events, as *Prisoner* fans do, or imagine props as *Game of Thrones* fans might, as they can go to one of these events instead, set up as the proper, official way to perform *Friends* fandom.

The ideal fan will therefore perform, but in ways that encourage further consumption of the brand: attending these sponsored events (which, it must be said, are not cheap), using the correct hashtags, and visiting the gift shop afterward. These all relate back to specific iconic moments that become representative of the show and can be continually resold in new forms. What is missing in these descriptions is opportunities for fans to perform outside of these iconic moments—to relive memories of the show outside of what they are provided, and to create and build upon these moments.

The Atomized *Friends* Fans

FriendsFest and the *Friends* pop-up are doing something right, at least according to the articles. FriendsFest, in its sixth year, largely sells out its weeks-long tour of the United Kingdom, while the *Friends* pop-up allegedly sold out its entire month-long run in a few hours. The idea that *Friends* fans want to go to an event centered around the show is clearly a justified one. There is a great deal of attachment to and affection for *Friends*, both in its home country and elsewhere, and this has been successfully channeled into events.

In this, *Friends* fandom potentially grows closer to more traditional fandom. As I talked about in previous chapters, place has an important role in building fandoms. Having a place concretizes what would otherwise be free-floating, giving a text a place in this world, rather than just on screen or in the mind. It also provides a space for fans to be fans, to perform and enact their fandom around others who get it. There is a power in being at the correct place for fandom. The connection made is more emotionally resonant than it is by performing elsewhere. This is one of the key factors of contemporary film and fan tourism, and one that seems to be successfully reproduced at FriendsFest and the *Friends* pop-up.

The promise of this connection draws fans to film places, potentially connecting them to other fans. This was appealing for the *Prisoner* fans in chapter 3, who tended to know fewer *Prisoner* fans in real life and enjoyed meeting them at PortmeiriCon, but also for the Harry Potter fans found in chapter 4, who enjoyed being in a space with so many other fans of the series, where they could "geek out" with impunity. Constructing the Wizarding World of Harry Potter or Portmeirion as the "place" of their respective fandoms involves connection not only with the text, but with the other people in fandom. Fandom has long been defined by its sociability, bringing together people with common interests to socialize and build upon their love of an original text. This social aspect is, to many, what makes fandom fandom, in the sense of an organized, coherent group. Place is often crucial to that, as it is through place that people can meet and get to know each other.

FriendsFest and the *Friends* pop-up are well situated to provide a similar place for *Friends* fans, especially as *Friends* is not considered part of the "cult" media landscape that is provided for by larger multifandom conventions. The only place that comes close to this function is the Warner Bros. studio tour in Los Angeles, and that must be shared with a variety of other sets and texts. However, what the *Friends* places promote is a more atomized experience, shared by the fans and those in their existing social networks who are also fans. There is little discussion of the gathering function of place, and instead a focus on, in the words of Comedy Central UK's Amanda Browne, "a great day out with your friends and family."[76] *Friends* fans, for all their volume, are conceived of as individuals or small groups, rather than a broader community as we might see at traditional fan conventions. In the texts, FriendsFest is not about meeting new *Friends* fans and seeing yourself as part of the community, but in being with *Friends* as a text.

This is particularly true for the pop-up. By design, it is structured for short trips—the FAQ on the website says to allow for thirty minutes to get through the entire space, and to leave time for the shop. As a series of rooms with props and reenactment opportunities, it is not set up for a meet-up with other fans. This tends to become clear in articles that feature tours through the space (as opposed to ones that simply announce the pop-up's existence), where the reporter and friends/colleagues are shown participating in the event, in the form of reenacting scenes and posing in the proper positions, but with little mention of other fans that are not brought along by the reporter. BuzzFeed's article is a good example of this: the staffers who attended are shown going through the space, posing at the correct moments, and ending on the street holding Central Perk to-go cups.[77] The experience is largely between the fan and the text of *Friends*, rather than the fan and other fans (with the exception of fans they already know, which, as I show in the previous chapter, can still be quite meaningful).

FriendsFest, as a space, is distinctly different than the *Friends* pop-up. As can be inferred from the name, the touring version (as compared with the 2015 London version, which was more similar to the New York pop-up) is conceived of as more of a festival. It is held outdoors in a park setting, and features food and drink stalls in addition to the larger set reconstructions and shops. It also boasts a large screen showing *Friends* clips, with an accompanying stage, something that is highlighted in photographs. Its FAQ suggests a two- to three-hour visit, in order to see and experience everything, pointing to its potential for lingering (and patronizing said food and drink stalls). That FriendsFest is larger and features more space for socializing between fans points to its potential to be the kind of gathering place for fandom that we see in the other chapters. However, this potential is rarely brought up. While a quiz is mentioned as part of the FriendsFest activities, usually briefly, the events that would be expected as part of a traditional fan convention, such as cosplay or panels, are missing. Activity is largely confined to the set tours and photo opportunities. Interestingly, it seems as if more involved fan events were once part of the planning for FriendsFest—an article from 2016 mentions that the stage would be used to "host a series of competitions, such as the Smelly Cat karaoke finalists."[78] This use of the space is not mentioned again, and by 2019, the focus is firmly on the re-creations, reenactments, and food options at the event. Even the quiz is downplayed—neither the Manchester *Evening News* nor MetroUK mentions it in previews of the event.

What this suggests is that the ideal form, and perhaps even the fan-preferred form, of fandom for *Friends* events is an atomized one, rather than a community-oriented or subcultural one. As I talk about a bit in chapter 1, there is a tension in fandom research between perspectives that look at fandom as a subculture and perspectives that look at it as an individual affective relationship with a particular text. In general, I've taken the latter approach here to fans and fandom, as it allows for a broader investigation of film tourism and those who do it. The affective aspects of fandom are, perhaps obviously, much bigger than the subcultural, and as I showed in previous chapters, film tourism has broad appeal. The emotional experience of "being there" with a favorite is appealing to many kinds of fans, whether they are involved with the subcultures of fandom or not.

However, tourism's popularity with non-subcultural fans is also due to the kind of trends I discuss in this chapter. "Being a fan" in an affective sense is a normalized position, a way of stressing that one deeply enjoys a text. In the contemporary media environment it bears very little stigma outside of whether other people you might encounter think the text is good or bad. Comparatively, subcultural fandom, while growing increasingly visible in popular culture, still bears some of the stigma of "excessive" behavior. The idea of "excessive fandom" is a continually negotiated one,[79] encompassing both subcultural and non-subcultural fans, and what is considered to be excessive behavior in fandom always depends on context. Yet it is clear that in fandom's mainstreaming, certain behaviors become more normal than others. While subcultural and community-oriented fandom might have liberatory potential, it is also not seen as normal in a way that just "loving" a show is. It is still where the "weirdos" are.

There is also an encouragement to reflect on *Friends* in a nostalgic, affective fashion. Lizardi[80] argues that contemporary mediated nostalgia is individuated, rather than collective, encouraging viewers to look inward rather than think about how their past connects with others. We are encouraged to focus on ourselves when we remember, a stance that is reflected in the way nostalgia is used in the marketing of FriendsFest and the pop-up. To return to the Maggie Wheeler quote I highlighted earlier, there is a recognition that *Friends* has played an important role in fans' lives, "through good times and bad times, through heartbreak and sickness, and through all kinds of life chapters."[81] Recalling *Friends* nostalgically through iconic moments and authentic props encourages an affective recollection of these moments, remembering when *Friends* has been there for the fans and contributed to their lives, furthering their loyalty to the brand.[82] The presence of props from filming also encourages a particular kind of reverence. They prove that *Friends* was important enough to remember and save in this fashion.[83] It is a complete work of art that should be celebrated as it is, rather than changed. Fans are encouraged to return to the series and remember moments that were important to them, and to link this to further introspective consumption, while leaving the text as it is.

If we think of FriendsFest and the *Friends* pop-up, and their accompanying texts, as designing an organized fandom for *Friends*, it is clear that the model being encouraged is atomized and affective, rather than subcultural. In comparison to subcultural fandom, where strong bonds are made with other fans and their fan works, strong bonds are made with the text in a way that keeps it under producer control. Busse reminds us that it is subcultural fandom that can best fit Jenkins's idea of fandom as resistant to dominant norms and the media industry, through building these bonds with other fans and through imagining and creating alternative versions of texts.[84] As I show in chapter 3, the bonds between fans can grow quite strong indeed, with the potential of overshadowing the fan's relationship with the text. For many *Prisoner* fans, it is meeting with fellow *Prisoner* fans, rather than watching the show or buying its merchandise, that keeps them going to Portmeirion. The subcultural bonds are long-lasting, contributing to the endurance of *Prisoner* fandom despite the lack of new material, but also potentially unruly. Fans as a group and a subculture have potential for co-opting the text in a way that creates a power struggle between them and the producers for meaning and ownership. While they might be devoted, this devotion comes at a price.

This can be seen with the other example of an industry-created fan place I explore, that of the Wizarding World of Harry Potter in the previous chapter. The experience at the Wizarding World is similarly atomized—made up of small groups of friends and family and centered around experiencing a deeper connection to the text. While fans appreciated having other fans around, and feeling part of the larger fandom, they mostly experienced it in an atomized fashion even though subcultural Harry Potter fandom is quite strong. Universal Studios itself has gradually phased out the more subcultural events it once hosted and positioned the park more toward engagement with the text itself. While the subcultural community is still a part of the park's audience, there is no longer an encouragement for them to gather as the fan community. Rather, it is friends-and-family visits that are encouraged, just like at *Friends* events.

In designing a fan experience, *Friends* has seemingly opted for the atomized model, which is already normalized, rather than encouraging the more participatory, subcultural kind of fandom we might see when fans are left to their own devices. The connections that fans should make there are with *Friends*.

Conclusion

What kind of fandom is encouraged by corporate places of fandom? That is the question I investigated in this chapter, through an analysis of the press and promotion around the UK's FriendsFest and the *Friends* pop-up in New York City. *Friends*, arguably the most popular show of the 1990s, has seen sustained interest in the 2010s and beyond, not only keeping many of its viewers from previous decades but gaining a large following of young people who discover the show via reruns or streaming. This renewed interest comes at a time when the media industry has taken an interest in fans and fan practices as a way to build "brand" loyalty to particular texts. While when it originally aired *Friends* was often used as the alternative to the "cult" texts that traditional organized fandom was built on, more recently there has been a recognition that *Friends* as a text encourages fannish attachment, and that many people do, in fact, consider themselves *Friends* fans. Throughout the past decade, the owners of *Friends* have worked to develop this fandom and brand loyalty toward the program.

One such strategy has been around places. As shown in the rest of this book, place is an important part of contemporary fandom, one that connects the fan to the text on a physical level and commemorates their affective involvement with it. Visiting places also provides the sort of "experience" that contemporary consumers are said to want from their favorite brands and products. As chapter 4 shows, it is possible for a re-created place to be embraced by fans, even without the aura of filming, and this mantle is picked up by these *Friends* places, both of which feature interactive re-creations of important *Friends* sites. Their popularity, alongside the lack of previous organized fandom for *Friends* compared with other corporate sites of fandom, makes them particularly interesting locations in which to investigate what kind of fandom is encouraged as film tourism moves from a grassroots to a corporate endeavor.

This chapter is therefore a complement to the previous chapters, showing how the developments I explore are being interpreted and used by the media industry. I do this through an analysis of the press and promotion around these events, situating the construction of ideal audience identity and behavior in the "hype" that prefigures the experience. In comparison with the tourist focus of the previous chapter, this one looks at how this experience is designed and idealized by the people who put it together. Through a rhetorical analysis of professionally produced articles about the events and supporting materials, I show how *Friends* fans and fandom are shown as normal, performative, and individualized.

First, *Friends* fandom is presented as a normal thing. There is no sense of stigma around being a fan of *Friends* that might want to attend FriendsFest or the pop-up, not even in a playful sense, as might be found with more traditional fan-oriented media. Rather, the language used to talk about *Friends* fans and their fandom is positive and inclusive, stressing not only that this is an event for *Friends* fans, but that "we" are all *Friends* fans. Behavior that might be considered extreme in other cases—such as proposing in Monica's re-created apartment—is treated positively and with enthusiasm, and when reporters show themselves at either FriendsFest or the pop-up, they are largely enthusiastic about the experience. This sense of *Friends* fandom—and visiting a *Friends* space—as normal is enhanced by the presence of celebrities at some of the events, many of whom display their own fandom toward the show while there. Being there is not depicted as a strange thing that needs to be explained to the audience, but as a desirable and normal experience.

This fandom is then expressed through performance. To be a fan in these settings, you have to do something—move around the spaces provided, take photographs, and reenact specifically selected moments from *Friends*. This suggests an understanding of fandom based around the idea that fans do things—that they are active and desire to do "more" than just watch television. Reenactment and performances are part of film tourism, expected as a way to mark out a space as "fan space" as well as connect the tourist with the narrative through use of their own body. The *Friends* places are, in many ways, quite similar. However, that these moments are fairly similar between FriendsFest and the *Friends* pop-up, despite being run in different countries by different companies, suggests a consolidating of particular moments as "iconic." Performing them demonstrates that one is "really" a fan of *Friends*, as they are what a fan should want to do, but also further strengthens particular aspects of *Friends* as part of the brand, solidifying its meanings and how fans should connect to and remember them. These are not places where fans are meant to wander and potentially make up their own stories.

Finally, this is done on an individual or small-group basis. While fandom has often been thought of as a social activity, bringing people together based on their shared interest, the kind of fandom promoted at FriendsFest and the *Friends* pop-up is more atomized. Attendees are invited to participate and interact with the re-creations alongside the friends they've brought along, but there is little mention of a broader fandom to meet with. The pop-up is especially designed for short encounters with the text, and while FriendsFest has potential to be a gathering place like those seen in previous chapters, the experience is promoted as an individual rather than a community one. Fans are encouraged to reflect on their fandom and what *Friends* has meant to them personally rather than thinking of themselves as a community. The ideal relationship is between the fan and the text, rather than the fan and other fans.

What does this suggest for the future of fandom and place? For one, the connection is likely to continue. The influence of the "experience economy" concept shows no sign of fading, despite recent developments, and place is one of the more credible ways to make an experience. The success of the *Friends* events, and of grander environments like the Wizarding World of Harry Potter, shows that there is an audience for it. However, the success of these events also shows how place can be used as a strategy of fan management. Through easy accessibility and the promise of an experience that can't be found elsewhere, fans are encouraged to participate in these controlled environments. With the creator in control of the space, which is presented as a gift or treat to fans, attention is turned back to the text and the producer, rather than the fan community and its own works. The fan, while having an interactive experience, is ultimately a passive participant in the text and story itself. The fan experience itself is removed from public space and commodified further. To have the "ultimate fan experience," you have to be willing to pay and stay within the boundaries of appropriate behavior.

As I showed in chapter 4, this can still be an emotionally meaningful, and very enjoyable, experience for fans. The media industry's resources mean that media companies can do things that fans can't, in terms of reproduction of environments, the amount and type of space they can acquire, or access to relics of filming. In many ways, creating environments like these is what fans want, and fans are happy to accept it. It must be noted that in this chapter's study I did not ask the fans themselves what they do in these places—it is possible, as we see in the Wizarding World of Harry Potter, that fans will appropriate these places to their own ends, regardless of how the sites are presented to them. But this subtle shift toward industry control through place is worth pointing out.

This book has shown that tourism connected to fandom is a powerful experience, as well as one growing in popularity. As tourism begins to take more prominence among other fan practices, it is worthwhile, however, to consider how these experiences might also be used as a method of fan control, directing fandom in a more docile, industry-friendly direction. This chapter is therefore a way of reflecting on some of the consequences, intended or unintended, of what we saw in the previous three chapters.

Conclusion
Film Tourism and the Future

Why do people engage in film tourism? It's not a scientific research question, to be sure, but it was the question I had in my head as I started this research project. I myself had never been particularly interested in visiting filming locations, despite my love of particular shows or movies—but others were, many others, and that fascinated me. What drew them there? What did people get out of it? These questions were in the back of my mind throughout the four and a half years I spent researching film tourism.

Here, I want to reflect on what those answers are, and what this might mean for the future of film tourism (a future that looked very different when I began writing). To do so, I will return to the three "modes" of imaginative experience in film tourism that I developed in chapter 2, using them as a starting point to think about the ways that fans experience film places, and, through that experience, how place changes the ways in which we think about fandom. From there, I will reflect on what contemporary developments might mean for film tourism into the future—whatever that future might be. In doing so, I illustrate what has been learned about film tourism in the researching and writing of this book, and highlight what work still needs to be done.

The Experience of Film Tourism

Film tourism, at least the form I talk about here, is a combination of two things—tourism and fandom. They both bring their own discourses and meanings to the experience, which combine to make film tourism what it is.

Without both these parts, film tourism would not be the growing phenomenon that it is.

Tourism is considered integral to modern life.[1] Undertaken for relaxation, education, and pleasure, "getting away" and seeing the sights of elsewhere is an idea that is almost uniformly popular. It also has a long history, from religious pilgrimages to the Great Tour of young aristocrats. People traveled to learn about other cultures and encounter the great works and sights of elsewhere, something that was considered important for education as well as for personal, Romantic fulfillment. As travel became less expensive and the idea of "educating" ordinary people took hold, tourism became a mass practice, not just an elite or special one. To show that you were a modern, enlightened person, you had to travel and see what were considered important, meaningful sites. This expectation has only intensified over the years. Tourism and travel are today somewhere between a right and a dream, emblematic of modern leisure practices and what a person might consider a lifelong goal.

Much of tourism is built around sightseeing. Seeing the important sights—the natural views, the great works of art and architecture, the historical locations—is emblematic of tourism, what one is supposed to do when one is somewhere else. This idea is consistently portrayed and reinforced in media of all kinds, from early guidebooks and travelogues to television and glossy magazines, and now on social media. Seeing important, beautiful, and famous sites is expected when one is traveling. These encounters are supposed to enlighten and improve the lives of tourists through experiencing something meaningful and valuable.[2]

Yet, just seeing is not enough. If it were, travel media would obliterate the need for physical travel—after all, one can, particularly these days, see nearly every important site in great detail and precision through the media. To "really" see something, one must be co-present with it,[3] to see it with one's own eyes. Without the presence of the body, it doesn't count as real. After all, we have grown used to the eye being "tricked" through media. To confirm the reality of something, and to have a genuine, meaningful encounter with it, one has to experience it through one's own body. The promise of tourism is to make these fantastic, unique things real on a personal, embodied level.

It is here that fandom comes in. While classic sights are valued by the broader culture, they might not have the intrinsic, personal value to a tourist that makes them worth visiting. What does have that value, increasingly, is film and television. Fans of movies and television shows have a bond with them. While fictional and created, these texts are, in important ways, real to fans. They are worlds and characters that the fans have thought about, contemplated, and internalized. They connect to what a fan considers personally meaningful and are a way of forming both an individual and group identity in modern society.[4] Essentially, objects of fandom matter on a deeply personal level.

It is this mattering that makes them a suitable focus for tourism and sightseeing. Visiting holds the promise of making these mediated images and narratives real—just as it does for more traditional forms of tourism. This is not to say that fans are delusional, thinking that the fictional narratives they love can be actually visited, but that seeing where the narratives "took place" gives them physical presence in the world. It allows for a play of boundaries between real and fictional in a similar, but differently felt, way as deep thinking about a narrative world. This can be thought of in multiple ways, as filmed texts like the ones I look at here always exist on multiple levels. While they are fictional, without any real-world counterparts, they are still made by people in our reality and leave traces in it that can be acknowledged and understood through tourism. Tourism, after all, confirms that something happened in a place, something worth being co-present with.

Within this focus, though, there are distinctions. Through the cases in this book, I developed and explored ways of understanding the variations in the film tourist experience. The three modes I settled on, and will reflect on again here, are not typologies, in the strictest sense, as they are not meant to delineate types of tourists. Rather, they are modes of experience that an individual fan tourist can slip in and out of during a visit. They illustrate the ways that fans make sense of the places they encounter, and the ways that fans use place to shape their fandom. The hyperdiegetic, production, and historical modes of film tourism reoccur throughout the four case studies in this book in different ways. Reassessing them here is a chance to reflect on the way we think about fans' relationship to place, and what the result of that can be.

The Hyperdiegetic Mode of Film Tourism

The hyperdiegetic mode of film tourism is the most "classic" form of film tourism: the fan tourist imagining that they are in the narrative through being at the place where it happens. Through visiting the place, tourists can imagine themselves as participating in the story but also can explore the narrative world through their own body, seeing and experiencing what might not be on screen but is nevertheless part of the story-world. Conceptualizing film tourism as this kind of immersion fantasy is found throughout the different disciplinary approaches toward film tourism, from Roesch's idea of "place insiderness,"[5] where tourists feel what characters felt, to Hills's idea that visiting places "sustain[s] cult fans' fantasies of 'entering' into the cult text."[6] Visiting places is the way to become part of the text. This is perhaps best represented by the tendency of all kinds of film tourists to reenact scenes at their appropriate locations. In reenacting the scenes, film tourists put themselves bodily in the place of the characters, sensorily experiencing what the characters experience. They become part of the story.

Game of Thrones fans explored in chapter 2 certainly reenacted—from large "ceremonies" honoring Robb Stark as the King in the North in Northern Ireland to fans walking down the Jesuit Stairs in Dubrovnik in the manner of Cersei Lannister (minus, at least in what I saw, the nakedness). Whenever fans were present, they would reenact the scenes in some fashion, remembering what happened there through their own performance. Even when fans were not openly performing, the story-world came in and out of their mind as they walked through the streets of old Dubrovnik or the coasts of Northern Ireland. In walking through and physically experiencing these places, they could develop their understanding of this world, freshly entered into their lives. What did the buildings of King's Landing feel like under their fingertips? What did the air of Winterfell smell like? These questions can be answered, in some way, through film tourism.

It also just felt extremely cool to "be there," to stand in the place where something important happened. As one interviewee remarked, "oh my God, that's King's Landing! We swam right next to it! That's fantastic!" *Game of Thrones* was new and exciting to these fans, and being able to stand—or swim—in the same place as the narrative was thrilling. It was a novel way to feel connected to the text, to learn about it, and bring that knowledge and experience back home. With new seasons beginning, it would enhance their viewing of the show—after all, they knew what the places were "really" like, beyond the audiovisual. They had been there.

For fans of *The Prisoner*, profiled in chapter 3, neither the text nor Portmeirion was new. *The Prisoner* was a completed text by 1968, before many of my interviewees had been born. Some had grown up with the text, and some had come to it in adolescence or young adulthood, but they were all very familiar with it by the time of the interviews. They were well aware of how it felt to be in the spaces of the narrative as they had usually spent a great deal of time in Portmeirion. It did not hold the same kind of mystery as Dubrovnik or Northern Ireland did for the interviewed *Game of Thrones* fans. Fans of *The Prisoner* also knew Portmeirion very well as Portmeirion itself, with its own narratives and experiences outside of the show. It was a home to them as much as a filming location.

Yet, despite this familiarity with the space in its own right, there was still a hyperdiegetic imagination in *Prisoner* fans' experience with the place. It was expressed in language used in describing different parts of Portmeirion—for example, in discussing Number 6's house, which was always referred to as such, or even describing the whole area as the Village, much as it was on the show. While *The Prisoner* wasn't everything to the fans interviewed here, it was still an integral part of how they understood the space. The show's narrative gave the places their first meanings, onto which other experiences were layered over time. The narrative was always there, ready to be remembered and recovered.

Reenactments were key to this memory. At PortmeiriCon, the narrative space was remembered and reexperienced through reenactments both small and large. Number 6's election campaign, the human chess game, and many smaller scenes were redone in the space they were filmed in, with parts assigned and lines rehearsed for accuracy. This reenactment gives participants the feeling that they are taking part in the narrative—they are following what series creator Patrick McGoohan (and several generations of fans) did, putting themselves in the mindset of the scene. Even after many years of doing so, it can be a powerful experience for convention-goers. There is still meaning in surrounding oneself with the text.

Indeed, that immersing oneself in the text is a meaningful experience to fans is the promise of the created environments I examine in this book, the Wizarding World of Harry Potter in chapter 4 and the two *Friends* events, FriendsFest and the *Friends* pop-up, in chapter 5. For fans of the Harry Potter series interviewed here, the Wizarding World of Harry Potter does, in fact, deliver on this hyperdiegetic immersion. Fans enjoyed feeling like they were part of the story-world and praised the construction of the park for the way that it created this atmosphere. Everything from sound to look to smell were praised by the interviewees, who, as fans of the series, felt qualified to evaluate its authenticity in relation to the narrative world being portrayed. The Wizarding World succeeded with these fans because all of it seemed designed to facilitate the immersive experience promised. It was a physicalization of the ironic imagination, the concept of treating the story-world as if it were real. Rides, shops, clothing, and even other park attendees all made the space feel like a living wizarding neighborhood, one that fans could imagine a space for themselves in. Fans talked about how they developed their own in-universe characters to walk around the Wizarding World, using the space to tell their own personal stories in the environment. Being there felt like entering into their fantasy.

In this, the Wizarding World of Harry Potter fulfilled the promise of the theme park—to be a set-apart space where the world outside of the park doesn't matter and the tourist can focus entirely on the fantasy. For Harry Potter fans, compared with Disney fans or more general theme-park fans, this meant a focus on the story-world of Harry Potter, separated not only from the rest of the world but from the rest of the park. The level of theming is seen as higher than at other parks or themed environments, and because of that level, fans were willing to participate in the theme and embrace it. While the Harry Potter franchise has more auratic locations, such as the film studio outside of London or the Platform 9¾ re-creation at King's Cross station, which are where the films were made and a named location in the books, respectively, the theme park is seen by fans as equally valid as a way to visit the story. It is the immersion fantasy they were promised.

This power of immersion into a narrative, and the belief that this is what fans want, was the focal point of the marketing of FriendsFest and the *Friends* pop-up. Both of these events centered around the 1990s sitcom *Friends* and sold themselves on a high level of immersion into it. As with the Wizarding World of Harry Potter, important locations were reproduced, with fans encouraged to inspect them for small details that would confirm their authenticity as reproductions. Fans could wander through them (in a supervised fashion), take pictures of themselves on the furniture, and re-create favorite scenes in the "proper" fashion. The promise of immersion into *Friends* and being able to explore the hyperdiegesis of the show was the core of the way they were marketed. It is assumed that this is what fans want out of their spatial experiences—a way to feel that they are part of the story-world. The fact that these events, like the Wizarding World of Harry Potter, are not auratic locations (although the pop-up is in New York, the setting of *Friends*) doesn't entirely matter. The promise of accurate immersion, of being able to do and see things that fans have wanted to do and see through years of viewing, is enough. Both FriendsFest and the pop-up have been successful, with the pop-up selling out for the month it was open and FriendsFest entering its sixth year, suggesting that the makers of these experiences are not wrong.

Immersion into the story-world is therefore an important part of the experience of film tourism, seen across the four cases analyzed here. However, this is not the only kind of imaginative experience that fans have with places. In Dubrovnik, for example, immersion into the narrative was fleeting—a moment here and there, rather than a complete blanketing of Dubrovnik with the show. *Prisoner* fans have developed a much deeper relationship with Portmeirion than just its use to encounter a story-world. Even fans at the Wizarding World of Harry Potter, which is a fully designed hyperdiegetic environment, managed to find other aspects of the place to highlight in interviews. Immersion is certainly part of what the fan experiences and seeks out at locations, but there is more to it.

The focus on immersion as the determining factor of why fans visit places and how they relate to them on site therefore limits our understanding of film tourism. It overshadows not only other ways of imaginatively experiencing a film-related place, as I will get to, but also other reasons why fans visit places to begin with. For example, reenactments can be read not only as a way of inserting oneself into a story-world but as a way of paying tribute to it. Fans reenact scenes to commemorate what happened there, both in the text and in real life. A *Prisoner* fan leading a reenactment thought of it as connecting her not only to Number 6 as a character, but to Patrick McGoohan as the series creator and especially to the fans that had come before her. Participating in the reenactment says that something that happened there mattered to her and to other fans. It connects her to the past and to the community of other fans, both present and imagined. Reenactments can be as much about commemorating texts as about imagining oneself in them.

Focusing on the fan connection to the story-world of a text rather than the other ways fans relate to them is part of a fan studies approach that generally privileges textual output. Fan fiction and other forms of narrative-based fan works are highlighted as the way that fans make sense of texts. The focus is on the story-world, which is perceived and marketed as the part that matters, particularly as fan texts turn into franchises that have multiple authorial voices and media forms. In a franchise environment, the story-world as a whole, rather than how it is depicted, is what matters. Fans appreciate franchises and spend a lot of time and effort developing and investigating them, but they are far from the only thing that fans enjoy about favorite texts. Seeing fandom as strictly textual limits our understanding of the ways that fans relate to fan objects just as much as it limits our understanding of how fans relate to places. Imagining the narrative, and developing one's own place in it, is important in the imaginative experience of film tourism, but it is not the only way.

The Production Mode of Film Tourism

As the interviews here—as well as previous research—show, fans are also interested in how favorite filmed texts came to be. Film tourism is not only about the narrative world of the text but about this kind of behind-the-scenes experience. This is one of the particularities of film tourism compared with other kinds of media tourism. Film exists on two levels: the world of the narrative and the world of the production, where the film is created through the labor of directors, set designers, actors, and so forth. Fantasies have to be physically made in a way that leaves a trace in the real world. Seeking out and understanding these traces is as much a part of film tourism as immersing oneself in the story-world.

We can see this clearly in the *Game of Thrones* tourist interviews in chapter 2. Fans at the various sites were very interested in how the show was created, particularly as the production altered so much of the original landscapes of both Dubrovnik and Northern Ireland to create Westeros. Fans were fascinated by how the production company found more obscure places, such as a storehouse in Lokrum, and used them to create a "realistic" mythic-medieval setting. This was seen across the different kinds of fans, whether they were involved in organized fandom for the show or not. The way the illusion was put together is just as interesting as the illusion itself. Understanding this process enhanced their appreciation of the story-world, rather than damaging it. Seeing the effort that was put into its creation made it feel more special than it did before.

This effect was capitalized on by tour guides in Dubrovnik, who provided a great deal of information about the production and "gossip" about the actors based on their experience as extras on the show and locals to the area. They could discuss how the production functioned on a day-to-day level, encouraging tourists to imagine what the making of the show was actually like. They could also discuss their personal experience with living in the place where the filming took place, and how the actors and crew made sense of the actual space of Dubrovnik, fostering a sense of imaginative inclusion in the production as well as with Dubrovnik itself. This kind of behind-the-scenes information was held up as a reason to take a paid tour, as the guides had much more information about the show than could be gleaned through a self-guided tour of the locations. Paid tours provided the tourists with more and better knowledge of the production, information that could enhance their viewing of the show and that they could bring back to show off to others. Imagining the production was as much of the experience as imagining being in the story.

For fans of *The Prisoner* in Portmeirion, the production mode of imagination is arguably more important than the hyperdiegetic. For this text, which is much less straightforward than *Game of Thrones* in terms of story-world, the motivations and mindset of series creator Patrick McGoohan plays a major role in fans' appreciation and imaginative engagement with it. Fans enjoy debating what McGoohan meant by the show and what messages can be taken from it. This kind of discussion makes up a great deal of fan chatter, both in person and in their writing over the years. In this discussion, Portmeirion itself plays a major role. It is credited with giving McGoohan the inspiration to create *The Prisoner*, as he visited it before he made the show, and giving the program its distinctive qualities. Portmeirion is not only the filming location of the show but an integral part of it, and fans feel that without Portmeirion, it would be a different, lesser program in terms of style and general uniqueness. Fans see Portmeirion as a paratext of *The Prisoner*, a way to better understand the show and how it was made.

Therefore, visiting Portmeirion and learning about it is more about accessing this kind of knowledge than it is about imagining oneself as part of the story-world. Being there allows fans to connect to McGoohan in both large and small ways, standing in spaces where he stood, experiencing the atmosphere of Portmeirion, and even staying in the cottage that he stayed in. Fans imagine themselves following in his footsteps. As Portmeirion itself is entwined with the origin story of *The Prisoner*, exploring it is also understanding how the show was created. Visiting therefore enhances one's *Prisoner* fandom by giving fans a better understanding and imaginative experience of why the show was the way it was, and what spatial qualities shaped it into being the text that they love.

Even fans at the Wizarding World of Harry Potter, a place that is structured entirely around the hyperdiegetic imagination, found ways to utilize the production mode. While the Wizarding World promises, and delivers, immersion into the narrative, fans know that this is the result of labor. This effort is widely appreciated. Fans frequently mention the work that went into making the Wizarding World good—the level of detail, the way that the detail was carried out, and so forth. Recognizing that the Wizarding World was created is part of appreciating it. The ironic imagination that enables one to experience the place "as if" it were real works because of the knowledge that it *isn't* real, that the fan is simply enjoying art for its own sake rather than being delusional. Seeing that a great deal of effort went into the production of the park enhances the fan's enjoyment of it.

Enjoyment of the park also becomes more culturally acceptable when the work to make it is recognized. Instead of just a theme park, the Wizarding World becomes a work of art and craft in its own right. One of the interviewees even became a Harry Potter fan because of how good the park was. Theme parks, for many, are seen as a medium in their own right, with their own formal qualities to evaluate. Appreciating them as created works can enable the immersive feeling that fans go there to experience, just as a well-made film is appreciated for its craft. Knowing that the Wizarding World of Harry Potter was created with skill means that it is a suitable place to immerse themselves in the fantasy, because it feels like a place where their fandom is valued.

As with *The Prisoner*, the author of the series also plays a role in fans' appreciation of the space. Fans paid attention to series creator J. K. Rowling's feelings about the park, and her endorsement of it was another factor in their appreciation of it. Knowing that she approved the elements of it, including the iconic food and drink, meant that it could be seen as an authorized, correct simulation of the story-world. Understanding the behind-the-scenes work that went into creating the Wizarding World establishes it as the correct place to live out hyperdiegetic fantasies. It is official, approved, and well constructed.

Behind-the-scenes information is also part of the marketing around FriendsFest and the *Friends* pop-up. While not as prevalent as the immersion fantasy, the promise of real props and the incorporation of actors from the show into its promotion also sought to acknowledge the draw of imagining the production of the show. Props are not just about giving the space the aura of legitimacy, although they are certainly part of it; they also offer the potential of gaining important knowledge. For a show like *Friends*, where fans have often watched the entire series a number of times, props represent one of the few chances to learn something new about it. They can see something about how the show was made, even if the focus of the events is imagining oneself in it.

What this shows is that the story-world is not the only draw of film tourism, and that behind-the-scenes information and experiences are also important to how fans make meaning out of film places. However, the experiences and information still must be connected to the text. Couldry suggests that an interest in behind-the-scenes media work is based in a fascination with the assumed specialness of media work, compared with other forms,[7] in transgressing the boundaries between the ordinary and the reified. Some evidence suggests that understanding how the medium works in general is part of the appeal, but, as Beeton's example of the failure of Fox Studios Australia demonstrates,[8] it is not enough to simply speak of media production in general. What is important to everyone interviewed here is the specific text that was made—that the location is where *Game of Thrones* or *The Prisoner* specifically was made. The true draw here is seeing how specific texts came to be, because those texts matter to fan tourists. There needs to be an existing connection for the fan to find meaning in understanding and imagining the process.

This desire for connection also indicates that we need to take a more expansive view of how fans relate to story-worlds. There is an assumption that fans have the most interest in the story-world, that their biggest interest as fans is in exploring its hyperdiegesis. The sort of "cult" texts that have pioneered fan practices are, after all, built around it,[9] and a great deal of emblematic fan practices involve further developing the story-world. Interest in film production is for a different kind of media consumer, the "film buff" who may not love any particular text but appreciates the formal qualities of the medium.

However, what this research shows is that even "affective" fans see, and appreciate, the creative work that goes into their favorite texts. They are interested in the way their favorites came to be and the work that goes into making them. They can have attachments to particular filmmakers or showrunners and think about how their qualities as artists shaped their work. There is not a strict separation between appreciating the world and appreciating the creation of it. While some fans might lean toward one or the other, it is more likely that, as with the modes of film tourism I present here, fans move in and out of different ways of appreciating texts. At one time, they are exploring the story-world and imagining new ways within it, and at another, they are discussing the way that the production team structured a particular shot or episode. It is all part of loving a favorite text.

The Historical Mode of Film Tourism

One of the concerns around film tourism is its relationship to existing place narratives and the impact that a filmic or textual identity would have on them.[10] Films and television shows are excellent storytellers, and their goal is to tell stories, not to necessarily showcase a place accurately or authentically. The fiction can easily dominate, as it is the narrative that viewers encounter first and become attached to. This issue is particularly complex when it comes to fantasies. Without any textual connection to the "real" landscape, there is potential for the existing place to get "lost" as fans rush in to give the story-world physical form.

However, in the case studies here, the relationship is much more complicated. Fan tourists display what I term a "historical mode" of imagination when visiting filmic places, a mode that draws them into the "real" histories and place narratives that surround the filming location. Fans were frequently interested in the places and landscapes around them and wanted to know more. Film tours were often used as a way to find new places and new narratives about them, ones that were not part of the "normal" tourist route. Once these tourists arrived, what happened textually and extratextually became ways into the unknown narratives of the place. Film tourism can therefore be a way to introduce these narratives and build new attachments to places.

This effect can be remarkably deep. As I show in chapter 3, many fans of *The Prisoner* consider themselves to be just as much fans of Portmeirion. *The Prisoner* brought them to the location, a holiday village designed by Sir Clough Williams-Ellis, but their long experience with it as individuals and as part of a bigger fandom has given it a meaning for them beyond the show. As frequent visitors, with many coming at least once a year for many years and some even more often than that, they came to know Portmeirion quite well. Their initial interest in how it inspired McGoohan to create *The Prisoner* shifted into an overall interest in how Portmeirion came to be and an appreciation of its own unique qualities as a place. They post and view pictures of Portmeirion on Facebook, debate new developments in the village, and even check in on the Portmeirion webcam when they need a boost. Many fans discussed in interviews how interested they became in the assembly of Portmeirion and Williams-Ellis's work as an architect through their *Prisoner* fandom. They could connect his story to McGoohan's use of Portmeirion as a sort of unintentional paratext, a way to better understand *The Prisoner*. Appreciating the place of filming went hand in hand with appreciating the show itself. After accepting it as the place of filming, fans developed their love of Portmeirion in its own right. It became another thing for them to be a fan of.

The fans here were also, as mentioned, frequent visitors to Portmeirion, either for the convention or for their own enjoyment outside of it. Through repeated visitation, they developed their own pathways and memories in Portmeirion that are connected to, but distinct from, *The Prisoner* itself. Portmeirion's long involvement with *Prisoner* fans, who have been coming yearly for conventions since the 1970s, sets the scene for this particular kind of place attachment. They can perform reenactments in the same place where they have been done for decades, stay in the same accommodation, and appreciate the same views. Visiting is less of a one-off "pilgrimage" than a homecoming, a return to a familiar fandom-related place. Fans can return, do things they've done before, track changes and new developments, and feel surrounded by their own fandom and that of the fans who have come before. At the same time, they love Portmeirion itself, and appreciate that it has a history outside of *The Prisoner*. It is that history that gives Portmeirion permanence. Regardless if the general public remembers *The Prisoner*, Portmeirion will remain, and they can go there and remember.

Game of Thrones fans showed an even more explicit interest in the history of the places they visited. This was brought up often in the interviews. Fans were very interested in the history of Dubrovnik and Northern Ireland and would often follow up on this interest when they went home. As with other kinds of tourists, who might be introduced to Dubrovnik or Northern Ireland through traditional promotional means, fan tourists want to learn about where they are. The show provided a frame for understanding, particularly as many fans were unfamiliar with these places and their narratives before visiting.

Game of Thrones itself worked in favor of this kind of imaginative experience. As a medieval fantasy, it put viewers into a historical mindset. Indeed, many were already interested in history before viewing the show, and only became more so as a result of their experience with it. Its promotional rhetoric, particularly at the time of my research, was about how much more accurate the show was in depicting the period than other medieval fantasies. While it was not about real places, fans nonetheless saw it as a realistic depiction of medieval Europe. Learning about the real places used in creating this specific fantasy helps reinforce the idea that *Game of Thrones* is a realistic depiction of what medieval life would have been like. It reinforces ideas about the show that its fans already held.

At the same time, it presents particular versions of the place narratives of Dubrovnik and Northern Ireland that fans were keen to pick up on and explore. As tourists, they were more interested in older historical sites, such as the Dark Hedges in Northern Ireland or the old center of Dubrovnik, than more contemporary ones involving the Troubles in Northern Ireland or the siege of Dubrovnik. This is not to say that they weren't interested—on a different day, they could be—but their particular interest in older narratives fit well with the *Game of Thrones* frame. The show's narratives were the way into imagining the histories of Dubrovnik and Northern Ireland. They shaped what fans wanted to see and how they reacted.

This phenomenon is seen in other studies of film tourism. If Vancouver is visited by an *X-Files* tourist, it becomes a place full of secrets waiting to be uncovered. A Harry Potter tourist in Britain searches for hidden magic. The links between the fictional and the historical narratives are therefore not new findings but should be emphasized here as part of the fandom focus this book takes. Fans already enjoy stories and demonstrate an ability and willingness to form affective attachments. Interest in one narrative can also lead to another, just as one fan's interest in a new text can bring in others. This is particularly the case when a show or film demonstrates similarity to a narrative they already know. Fans don't usually restrict themselves to one text. In tourism terms, interest in and attachment to the fictional narrative can be brought over to the historical narrative, particularly if connections are made between them. As we see with fans of *The Prisoner*, this can end up being a very strong attachment, one that exists in tandem with the text that brought them there. Fans don't necessarily want to have the text overshadow the actual place—many of them want both to exist, and to use the text as a frame to understand the place.

The historical experience can also shape the way the text is understood. *Prisoner* fans once again demonstrate the potential of this process. Understanding Portmeirion gives them new insight into *The Prisoner* by giving them more understanding of the kind of iconoclastic artistic temperament that makes such a work. Similarly, understanding real history as demonstrated in Dubrovnik and Northern Ireland shapes the understanding of how the world of *Game of Thrones* works. Fans can use this knowledge to expand on the narrative presented, filling in gaps and developing new theories. In combing narratives, both are enhanced.

This is missing in the final two case studies, that of the Wizarding World of Harry Potter and the *Friends* events in the United Kingdom and United States. They are both deliberately crafted to be "set apart" from reality and real places, with little connection between the narrative world and the physical place they happen to be in. In the case of the Wizarding World, this is enforced through the design of the place, with Universal Studios set apart from Orlando, and the Wizarding World further set apart from the rest of the park. Part of the medium specificity of theme parks is that they are apart from the world. Being outside of the "real world," they become a place to pretend, a place to imagine a story-world through physical experience with it. The outside world is irrelevant to the story being told within the enclosed space. At FriendsFest and the *Friends* pop-up, the separation is less intense, but still enacted. There are entrances, fences, and all the architecture of enclosure. Fans are expected to leave the real world behind and ignore that they are not in the "correct" location of filming, because what is within the enclosure is more important and interesting.

What matters is the more direct, intense connection to the fiction. The Wizarding World of Harry Potter and the *Friends* events both work because they shut off, to at least a certain degree, the historical mode of imagination. There is nothing to envision about the past, no outside narrative to connect to and build upon. Rather, there is a more complete connection to the text as it is made—a physical experience as close to the "real thing" as one can get, given official backing by the people who made the original. The fans here do not have the chance to make the connections that *Prisoner* fans have made with Portmeirion or that *Game of Thrones* fans make with Dubrovnik and Northern Ireland, but that is also not the point. There are places where those connections can be made for each of these texts. What is sought out in simulated places is the disconnection—the idea that, if just for a moment, there is only the fiction.

However, theme parks and other simulations are themselves places, set apart as they are, and this raises the possibility of a future historical connection. As Williams's work on theme park fans shows,[11] fans do get attached to the specifics of simulated locations and how they operate and develop, in a way that is not all that different from how *Prisoner* fans relate to Portmeirion. It is too early to tell, and a bit beyond the scope of this research, to say that this will happen with the Wizarding World of Harry Potter, but it is not outside the realm of possibility. Design can't stop visitors from building a history with a place and carrying that history into their interactions with it. As with the Disney fans analyzed by Williams, there is potential for fans at these simulated locations to build a new kind of historical imagination around them, different from what is experienced at locations with existing place narratives. With the Wizarding World in Orlando now more than a decade old, it would be worth exploring how fans' relationship with the park has changed, and what sort of new historical imaginations are made from simulated places as they proliferate.

The Future of Film Tourism

I started this book in a much different place than when I finished it. The field of film tourism research has grown up around me, particularly from the media and fan studies side, where so much exciting new work has emerged. New films and locations have risen to prominence, and the ones I first studied have begun to fade (ask anyone who follows the show about the *Game of Thrones* finale). More monumentally, when I turned in the first draft of this manuscript, I was just hearing that a new virus had reached Europe. Within two weeks of that submission, the entire tourism industry had changed as COVID-19 shut down borders and countries went into lockdown. As I write this, the pandemic is still ongoing, and the tourism industry is in disarray. An interesting time to be writing a book on film tourism, to say the least.

As I sit here, it is difficult to say what the future holds for tourism, much less film tourism. When this book is published, will we be gleefully off on trips again? Will the industry recover? Will the cycle of lockdowns and adjustments cause us to rethink what we travel for, or will time spent with our screens drive us to deeper connections with texts and a corresponding drive to visit their places of production? In these times, so much is unknown.

Yet the unpredictability of the future has always been part of studying media. What will and won't succeed is not determined by entering data points into a formula, even in our contemporary age of algorithm-informed media research. There are always variables when it comes to art and entertainment, things that can't be predicted, audience responses that can't be foreseen. As I always tell students, if the entertainment industry could predict with certainty what would be a success or failure, or if marketing could cover all artistic sins, there would be no flops. There is still some level of uncertainty anytime an audience gets involved.

For example, when I was writing the chapter in this book on the Wizarding World of Harry Potter, I, and most of the other people writing about theme parks, expected that Disney's then-upcoming *Star Wars*–themed area would at least match, and probably outpace, the Wizarding World's success. After all, *Star Wars* was just about the only media franchise that could match Harry Potter's popularity and had just returned to prominence with a successful reboot series. Combined with Disney's expertise in theme park design and construction, this seemed like a no-brainer. However, this doesn't seem to be the case. While Disney is characteristically tight-lipped about struggles, commentary suggests that Galaxy's Edge was relatively sparsely attended even before COVID-19.[12] While it is likely that Disney will iron out the issues as time goes on, it is interesting that what seemed like as sure of a bet as one can make in this field has turned out to be at least a mild disappointment. Even the Harry Potter franchise itself has struggled due to J. K. Rowling's transphobic comments and (perhaps much more of an issue for Warner Bros.) the failure of the second film of a planned *Fantastic Beasts and Where to Find Them* trilogy. Film tourism's main difference from regular tourism is its dependence on these kinds of outside narratives. If a film isn't successful and doesn't connect with an audience, then the location won't be appealing. Indeed, it is the connection to the audience that makes film tourism worthwhile. Fans of *Breaking Bad* visit Albuquerque not because the show makes it look appealing—it does not—but because they want to connect to and commemorate the show that has meant so much to them. Increasingly, the way to do that is through physical connection.

If there is one thing the COVID-19 experience has taught us, it's that co-presence still matters. Virtualization has grown substantially as a safe way to hold events, but at least anecdotally, it is not considered the same as being there. The desire for physical tangibility and co-presence I discuss as being at the heart of film tourism has not gone away. If anything, the lack of it just makes it more desirable. Many of us are looking forward to when we are able to travel and meet face-to-face again. Williams's discussion of theme park fandom in the age of COVID-19 shows that the desire, even the need, to be physically present in the place of fandom is a powerful one,[13] and fans have returned quickly as the parks reopened despite the potential risks.

This desire for tangibility in the face of digitalization and virtualization is a running theme not only in this book, but in fandoms of all kinds in recent years. Film tourism is only one example of the pre-COVID trend toward the physicalization of fandom. As I discuss in chapter 3, conventions are a foundational element of sci-fi and fantasy fandom, bringing fans together before there was a wide-ranging electronic infrastructure to do so. As with fannish practice more generally, this tradition has expanded, with everything from YouTube celebrities, Bravo reality television, RuPaul's *Drag Race*, and the "documentary" series *Ancient Aliens* all having their own. Despite the internet's promise to bring fans together and negate the need for a face-to-face gathering, the draw of physical connection to the fandom and its luminaries has proven resilient. There is power in physicality, especially as more and more of our daily experience of culture happens digitally, and this appeals to a much wider variety of fans than previously realized.

There is also an increased awareness of the potential value of these connections. Places and place-based events offer opportunity for media industries. In creating and managing these events and places, they can organize them in ways that benefit their interests: showcasing their interpretation of the text, promoting specific moments as iconic, and enforcing their status as the official source of information and engagement with the text. Rather than waiting for fans to visit and make meaning out of places for themselves, media companies are aware that place is a lucrative brand management opportunity, a chance to shape and guide fan behavior.

This trend is likely to continue into the future, and it will have an impact on fandom, both as a practice and as a subculture. Jenkins, in 1992, described fandom as a form of resistance, a sort of struggle for control over popular media between the audience and the media industry.[14] While the industry made the texts, fans interpreted them how they wanted, and built their own infrastructures around them without industry input. By 2006, there was a belief, from both him and others,[15] that new media could empower fandom, allowing fans to organize and create at almost the same level as the media industry. Fans could no longer be ignored or written off as weirdos—they were vocal and noticeable, and their transformative work promised an independent way of engaging with popular culture and challenging the dominant order.

However, as Booth[16] and Stanfill[17] show, this visibility can be turned back on fandom. When fans show what they want, they can be addressed in ways that bring control back to the industry, with the potential for challenge to their authority in defining and using the text minimized.

As we see in chapter 5, this is likely to be at least the short-term effect of the rise of film tourism. Place, especially, is something where the superior resources of the media industry compared with fan groups become evident. With the finances to create reproductions, control access to spaces of filming, and work with producers, the media industry has many advantages over smaller, more grassroots fan organizations when it comes to place and place-based experiences. While fans can hold a convention, with the better-resourced ones even hosting actors or other creatives, even the best-resourced fans can't pull off something like FriendsFest, much less the Wizarding World of Harry Potter. The level can't be matched. As a result, places of fandom have come to be controlled by the media industry, requiring certain amounts of money to access and adherence to certain norms. They focus on the individual fan and the fan's relationship with the text, which can be enhanced or strengthened by visiting these places and purchasing related experiences and merchandise. How these places develop and proliferate is a worthwhile focus of future research on place and fandom.

However, we must be wary of totalizing here. As the struggles of Galaxy's Edge show, there is still uncertainty as to what fans will or will not embrace, and what they will and won't put up with. The original idea of fandom as resistance also came at a time when the media industry's power felt complete, only for deeper ethnographic research to complicate this. Williams's work on fans of the Disney parks and the negotiations they make with the company are a clear example of the complications that are still uncovered in researching fan places, especially corporate ones. This is a promising and necessary direction for future research. The thing about fandom is that it is complex; trying to categorize it in the way that academics like to do is often a futile exercise. Film tourism showcases these contradictions well. At once grassroots and corporate, accessible and elitist, creative and passive, film tourism is hard to categorize in simple terms, which makes it hard to predict what its future form will take, even without a pandemic to take into account.

Whether this book is a direction for the future, or a snapshot of an era left behind after the pandemic, only time will tell. I hope that, regardless, it is useful in looking at how place, media, and fandom combine to make meaning. Whatever forms the future takes, I know we will continue to appreciate stories and the places they inhabit. It is that appreciation that shapes everything here.

Notes

Introduction

1. Nick Bramhill, "Force Awakens at Skellig Michael as Busiest Ever Tourism Season Begins," *Irish Independent*, May 8, 2017, https://www.independent.ie /irish-news/force-awakens-at-skellig-michael-as-busiest-ever-tourism-season -begins-35690097.html.

2. Chung Ah-young, "Descendants of the Sun' Makes Taekbaek a Tourism Magnet," *The Korea Times*, March 24, 2016, https://www.koreatimes.co.kr /www/news/nation/2016/03/116_201065.html.

3. John Urry and Jonas Larsen, *The Tourist Gaze 3.0* (London: Sage, 2011), 4–5, 155–189.

4. Nicola Watson, *The Literary Tourist: Readers and Places in Romantic & Victorian Britain* (Basingstoke: Palgrave Macmillan, 2006).

5. Sue Beeton, "The Advance of Film Tourism," *Tourism and Hospitality Planning & Development* 7, no. 1 (2010): 1–6.

6. Angelina I. Karpovich, "Theoretical Approaches to Film-Motivated Tourism," *Tourism and Hospitality Planning & Development* 7, no. 1 (2010): 7–20.

7. Joanne Connell, "Film Tourism—Evolution, Progress, and Prospects," *Tourism Management* 33, no. 5 (2012): 1007–1029.

8. Connell, "Film Tourism," 1009.

9. See Christine Lundberg and Vassilios Ziakas, eds., *The Routledge Handbook of Popular Culture and Tourism* (London: Routledge, 2019); Maria Månsson, Annæ Buchmann, Cecilia Cassinger, and Lena Eskilsson, eds., *The Routledge Companion to Media and Tourism* (London: Routledge, 2021).

10. See *The Journal of Popular Culture* 52, no. 6, "Exploring the Popular Culture and Tourism Place Making Nexus"; *JOMEC Journal* 14, "Transmedia Tourism."

11. Anne Marit Waade, "Mind the Gap—Interdisciplinary Approaches to Media and Tourism," in *The Routledge Companion to Media and Tourism*, ed. Maria Månsson, Annæ Buchmann, Cecilia Cassinger, and Lena Eskilsson (London: Routledge, 2021), 24.

12. See Paul Booth, *Playing Fans: Negotiating Fandom and Media in the Digital Age* (Iowa City: University of Iowa Press, 2015); Matt Hills, *Fan Cultures* (London: Routledge, 2002); Katherine Larsen and Lynn Zubernis, *Fandom at the Crossroads: Celebration, Shame and Fan/Producer Relationships* (Newcastle upon Tyne: Cambridge Scholars Publishing, 2012); Cornel Sandvoss, *Fans: The Mirror of Consumption* (Cambridge: Polity, 2005).

13. Connell, "Film Tourism," 1009.

14. See Joanne Connell and Denny Meyer, "Balamory Revisited: An Evaluation of the Screen Tourism Destination-Tourist Nexus," *Tourism Management* 30, no. 2 (2009): 194–207; Anita Fernandez-Young and Robert Young, "Measuring the Effects of Film and Television on Tourism to Screen Locations: A Theoretical and Empirical Perspective," *Journal of Travel & Tourism Marketing* 24, nos. 2–3 (2008): 195–212; Sangkyun Kim, "Extraordinary Experience: Re-enacting and Photographing at Screen Tourism Locations," *Tourism and Hospitality Planning & Development* 7, no. 1 (2010): 59–75.

15. Sue Beeton, *Film-Induced Tourism*, 2nd ed. (Bristol: Channel View Publications, 2016), 9.

16. Lincoln Geraghty, Vassilios Ziakas, and Christine Lundberg, "Guest Editorial: Exploring the Popular Culture and Tourism Place Making Nexus," *The Journal of Popular Culture* 52, no. 6 (2019): 1241–1249.

17. Stijn Reijnders, *Places of the Imagination: Media, Tourism, Culture* (Farnham: Ashgate Publishing, 2011), 5.

18. Watson, *The Literary Tourist*.

19. Leonieke Bolderman, *Contemporary Music Tourism: A Theory of Musical Topophilia* (London: Routledge, 2020); Cornel Sandvoss, "'I♥Ibiza': Music, Place, and Belonging," in *Popular Music Fandom: Identities, Roles, and Practices*, ed. Mark Duffet (New York: Routledge, 2014), 115–145.

20. Ross Garner, "Finding Nemo's Spaces: Defining and Exploring Transmedia Tourism," *JOMEC Journal* 14 (2019): 11–32.

21. Beeton, "The Advance of Film Tourism."

22. Karpovich, "Theoretical Approaches."

23. Connell, "Film Tourism."

24. Karpovich, "Theoretical Approaches," 11–15.

25. Dean MacCannell, *The Tourist: A New Theory of the Leisure Class*, 2nd ed. (Berkeley: University of California Press, 1999).

26. Connell, "Film Tourism."

27. Mel Stanfill, *Exploiting Fandom: How the Media Industry Seeks to Manipulate Fans* (Iowa City: University of Iowa Press, 2019).

28. Reijnders, *Places of the Imagination.*

29. Colin McGinn, *Mindsight: Image, Dream, Meaning* (Cambridge: Harvard University Press, 2004), 49.

30. Hills, *Fan Cultures*, 137.

31. Will Brooker, "The *Blade Runner* Experience: Pilgrimage and Liminal Space," in *The Blade Runner Experience: The Legacy of a Science-Fiction Classic*, ed. Will Brooker (London: Wallflower Press, 2005), 47–107; Hills, *Fan Cultures*, 110–121; Stefan Roesch, *The Experiences of Film Location Tourists* (Bristol: Channel View Publications, 2009), 114.

32. Nigel Thrift, *Spatial Formations* (London: Sage, 1996).

33. Lincoln Geraghty, *Cult Collectors: Nostalgia, Fandom, and Collecting Popular Culture* (London: Routledge, 2014).

34. Sabrina Mittermeier, "(Un)Conventional Voyages?—*Star Trek: The Cruise* and the Themed Cruise Experience," *The Journal of Popular Culture* 52, no. 6 (2019): 1372–1386.

35. Lynn Zubernis and Katherine Larsen, "Make Space for Us! Fandom in the Real World," in *A Companion to Media Fandom and Fan Studies*, ed. Paul Booth (Hoboken: Wiley Blackwell, 2018), 145–159.

36. Philipp Dominik Keidl, "Behind-the-(Museum)-Scenes: Fan-Curated Exhibitions as Tourist Attractions," in *The Routledge Companion to Media and Tourism*, ed. Maria Månsson, Annæ Buchmann, Cecilia Cassinger, and Lena Eskilsson (London: Routledge, 2021), 297–306; Dorus Hoebink, Stijn Reijnders, and Abby Waysdorf, "Exhibiting Fandom: A Museological Perspective," *Transformative Works and Cultures* 16 (2014).

37. Gilbert Rodman, *Elvis after Elvis: The Posthumous Career of a Living Legend* (London: Routledge, 1996), 97–122.

38. Hills, *Fan Cultures*, 110–121.

39. Sandvoss, *Fans*, 44–66.

40. Hills, *Fan Cultures*, 153.

41. Sarah Baker, Lauren Istvandity, and Raphaël Nowak, "The Sound of Music Heritage: Curating Popular Music in Music Museums and Exhibitions," *International Journal of Heritage Studies* 22, no. 1 (2015): 70–81; Keidl, "Behind-the-(Museum)-Scenes," 297–306; Marion Leonard, "Exhibiting Popular Music: Museum Audiences, Inclusion and Social History," *Journal of New Music Research* 39, no. 2 (2010): 171–181; Christian Hviid Mortensen, "Commemorating Popular Media Heritage: From Shrines of Fandom to Sites of Memory," in *The Routledge Companion to Media and Tourism*, ed. Maria Månsson, Annæ Buchmann, Cecilia Cassinger, and Lena Eskilsson (London: Routledge, 2021): 267–276.

42. Christoph Hennig, "Tourism: Enacting Modern Myths," trans. Alison Brown, in *The Tourist as a Metaphor of the Social World*, ed. Graham Dann (Oxon: CABI Publishing, 2002), 169–187.

43. Anthony Giddens, *Modernity and Self-Identity: Self and Society in the Late Modern Age* (Cambridge: Polity Press, 1991); Rebecca Williams, *Post-Object Fandom: Television, Identity, and Self-Narrative* (New York: Bloomsbury, 2015).

44. Melissa Beattie, "The '*Doctor Who* Experience' (2012–) and the Commodification of Cardiff Bay," in *New Dimensions of Doctor Who: Adventures in Space, Time, and Television*, ed. Matt Hills (London: I.B. Tauris, 2013), 177–191.

45. Booth, *Playing Fans*, 101–122.

46. Ross Garner, "Symbolic and Cued Immersion: Paratextual Framing Strategies on the *Doctor Who* Experience Walking Tour," *Popular Communication* 14, no. 2 (2016): 86–98.

47. Bethan Jones, "'*The Walking Dead* Family Is a Real Thing, Not Just a Hashtag': Theorizing Dissonance through Fan Tourism for *The Walking Dead* in Woodbury, Atlanta," *JOMEC Journal* 14 (2019): 53–70.

48. Urry and Larsen, *The Tourist Gaze 3.0.*

49. MacCannell, *The Tourist.*

50. Glen W. Croy, "Film Tourism: Sustained Economic Contributions to Destinations," *Worldwide Hospitality and Tourism Themes* 3, no. 2 (2011): 159–164.

51. Although see Beeton, *Film-Induced Tourism*, for a broader view.

52. Ashok Selvam, "Look Around the 'Saved by the Bell' Pop-Up, Debuting Today in Chicago." *Eater Chicago*, June 1, 2016, https://chicago.eater.com /2016/6/1/11827664/saved-by-the-bell-photos-gallery-menus-chicago.

Chapter 1

1. Urry and Larsen, *The Tourist Gaze 3.0*, 21–23.

2. David Crouch, "Places Around Us: Embodied Lay Geographies in Leisure and Tourism," *Leisure Studies* 19, no. 2 (2000): 63–76.

3. Hills, *Fan Cultures.*

4. Sandvoss, *Fans.*

5. Reijnders, *Places of the Imagination.*

6. See Sean Moores, *Media, Place, and Mobility* (Basingstoke: Palgrave Macmillan, 2012).

7. Connell, "Film Tourism," 1025.

8. Ibid.

9. Karpovich, "Theoretical Approaches," 7–18.

10. Robert Moses Peaslee, "One Ring, Many Circles: The Hobbiton Tour Experience and a Spatial Approach to Media Power," *Tourist Studies* 11, no. 1 (2010): 42.

11. Croy, "Film Tourism: Sustained Economic Contributions"; Francesco Di Cesare, Luca D'Angelo, and Gloria Rech, "Films and Tourism: Understanding

the Nature and Intensity of Their Cause–Effect Relationship," *Tourism Review International* 13, no. 2 (2009): 103–111.

12. Nelson H. Graburn, "The Anthropology of Tourism," *Annals of Tourism Research* 10, no. 1 (1983): 11–17.

13. Urry and Larsen, *The Tourist Gaze 3.0*, 5.

14. Ibid., 1.

15. Dean MacCannell, *The Ethics of Sightseeing* (Berkeley: University of California Press, 2011), 42.

16. See Daniel J. Boorstin's *The Image: A Guide to Pseudo-Events in America* (New York: Vintage Books, 1962) for an example of this kind of critique.

17. Judith Adler, "Origins of Sightseeing," *Annals of Tourism Research* 16, no. 1 (1989): 7–29.

18. Adler, "Origins of Sightseeing," 22.

19. Urry and Larsen, *The Tourist Gaze 3.0*, 5.

20. Ibid., 2.

21. Ibid., 18.

22. Ibid., 15.

23. MacCannell, *The Tourist*, 42–48.

24. Ibid., 45.

25. Urry and Larsen, *The Tourist Gaze 3.0*, 4.

26. Mike Crang, "Knowing, Tourism and Practices of Vision," in *Leisure/Tourism Geographies: Practices and Geographical Knowledge*, ed. David Crouch (London: Routledge, 1999): 238–257.

27. Olivia Jenkins, "Photography and Travel Brochures: The Circle of Representation," *Tourism Geographies* 5, no. 3 (2003): 305–328.

28. John Urry, *The Tourist Gaze*, 2nd ed. (London: Sage, 2002), 129.

29. Ibid., 19; see also Chris Rojek, "Indexing, Dragging, and the Social Construction of Tourist Sites," in *Touring Cultures: Transformations of Travel and Theory*, ed. Chris Rojek and John Urry (London: Routledge, 1997), 52–74.

30. André Jansson, "A Sense of Tourism: New Media and the Dialectic of Encapsulation/Decapsulation," *Tourist Studies* 7, no. 1 (2007): 5–24.

31. Maria Månsson, "Mediatized Tourism," *Annals of Tourism Research* 38, no. 4 (2011): 1634–1652.

32. Henrik Linden and Sara Linden, *Fans and Fan Cultures: Tourism, Consumerism and Social Media* (London: Palgrave Macmillan, 2017), 115–121.

33. Urry and Larsen, *The Tourist Gaze 3.0*, 21.

34. Tom Edensor, "Performing Tourism, Staging Tourism: (Re)Producing Tourist Space and Practice," *Tourist Studies* 1, no. 1 (2001): 59–81; Jonas Larsen, "Families Seen Sightseeing: Performativity of Tourist Photography," *Space and Culture* 8, no. 4 (2005): 416–434; Tijana Rakić and Donna

Chambers, "Rethinking the Consumption of Places," *Annals of Tourism Research* 39, no. 3 (2012): 1612–1633.

35. Crouch, "Places Around Us," 68.

36. Paul Rodaway, *Sensuous Geographies: Body, Sense, and Place* (New York: Routledge, 2002), 159–171.

37. Philip Auslander, *Liveness: Performance in a Mediatized Culture* (London: Routledge, 1999), 10–60.

38. John Fiske, "The Cultural Economy of Fandom," in *The Adoring Audience: Fan Culture and Popular Media*, ed. Lisa A. Lewis (London: Routledge, 1992), 42.

39. Beeton, *Film-Induced Tourism*, 41–45.

40. Nick Couldry, *The Place of Media Power: Pilgrims and Witnesses of the Media Age* (London: Routledge, 2000), 65–104.

41. Although see Maura Grady and Tony Magistrale, *The Shawshank Experience: Tracking the History of the World's Favorite Movie* (New York: Palgrave Macmillan, 2016); Christina Lee, "'Have Magic, Will Travel': Tourism and Harry Potter's United (Magical) Kingdom," *Tourist Studies* 12, no. 1 (2012): 52–69; Richard Roberson and Maura Grady, "The 'Shawshank Trail': A Cross Disciplinary Study in Film Induced Tourism and Fan Culture," *AlmaTourism* 6, no. 4 (2015): 47–66; and Reijnders, *Places of the Imagination*.

42. Roberson and Grady, "The 'Shawshank Trail,'" 48–51.

43. Sandvoss, *Fans*, 8.

44. Hills, *Fan Cultures*, 17–44.

45. Lawrence Grossberg, "Is There a Fan in the House? The Affective Sensibility of Fandom," in *The Adoring Audience: Fan Culture and Popular Media*, ed. Lisa A. Lewis (London: Routledge, 1992): 50–65.

46. Kristina Busse and Jonathan Gray, "Fan Cultures and Fan Communities," in *The Handbook of Media Audiences*, ed. Virginia Nightingale (Chichester: Wiley Blackwell, 2011), 425–443.

47. Leshu Torchin, "Location, Location, Location: The Destination of the Manhattan TV Tour," *Tourist Studies* 2, no. 3 (2002): 247–266.

48. Beeton, *Film-Induced Tourism*, 225–235.

49. But not always: see Daniela Carl, Sara Kindon, and Karen Smith, "Tourists' Experiences of Film Locations: New Zealand as 'Middle-Earth,'" *Tourism Geographies* 9, no. 1 (2007): 49–63.

50. Grossberg, "Is There A Fan?," 50–65.

51. Hills, *Fan Cultures*, 105.

52. Henry Jenkins, *Textual Poachers: Television Fans and Participatory Culture* (New York: Routledge, 1992), 51–87.

53. Michael Saler, *As-If: Modern Enchantment and the Literary Prehistory of Virtual Reality* (New York: Oxford University Press, 2012), 30.

54. Reijnders, *Places of the Imagination*, 15–20.

55. Ibid., 15.

56. Sandvoss, *Fans*, 61.

57. Hills, *Fan Cultures*, 150.

58. Roesch, *The Experiences*, 114.

59. Reijnders, *Places of the Imagination*, 55–80.

60. Rodanthi Tzanelli and Majid Yar, "*Breaking Bad*, Making Good: Notes on a Televisual Tourist Industry," *Mobilities* 11, no. 2 (2016): 188–206.

61. Couldry, *Media Power*, 75.

62. Reijnders, *Places of the Imagination*, 14.

63. Hills, *Fan Cultures*, 149.

64. Williams, *Post-Object Fandom*, 21.

65. Hills, *Fan Cultures*, 149.

66. Beeton, *Film-Induced Tourism*, 43–48.

67. Anne Buchmann, Kevin Moore, and David Fisher, "Experiencing Film Tourism: Authenticity & Fellowship," *Annals of Tourism Research* 37, no. 1 (2010): 229–248.

68. Sandvoss, *Fans*, 63.

69. Reijnders, *Places of the Imagination*, 104–106.

70. Ellen Badone and Sharon R. Roseman, eds., *Intersecting Journeys: The Anthropology of Pilgrimage and Tourism* (Urbana: University of Illinois Press, 2004); Urry and Larsen, *The Tourist Gaze 3.0*.

71. Rodaway, *Sensuous Geographies*, 41–60.

72. Reijnders, *Places of the Imagination*, 17–19.

73. Erik Cohen, "Authenticity and Commoditization in Tourism," *Annals of Tourism Research* 15, no. 3 (1988): 371–386; Ning Wang, "Rethinking Authenticity in Tourism Experience," *Annals of Tourism Research* 26, no. 2 (1999): 349–370.

74. Marie-Laure Ryan, *Narrative as Virtual Reality: Immersion and Interactivity in Literature and Electronic Media* (Baltimore: Johns Hopkins University Press, 2001), 39.

75. Jay David Bolter and Richard Grusin, *Remediation: Understanding New Media* (Cambridge: MIT Press, 1999); Oliver Grau, *Virtual Art: From Illusion to Immersion*, trans. Gloria Custance (Cambridge: MIT Press, 2003).

76. Brian Attebery, *Strategies of Fantasy* (Bloomington: Indiana University Press, 1992).

77. Katherine A. Fowkes, *The Fantasy Film* (Chichester: Wiley-Blackwell, 2010), 5.

78. Attebery, *Strategies of Fantasy*.

79. Ibid.

80. Fowkes, *The Fantasy Film*, 5.

81. Catherine Johnson, *Telefantasy* (London: British Film Institute, 2005), 7.

82. Rodaway, *Sensuous Geographies*, 160–163.

83. Johnson, *Telefantasy*; Jacqueline Furby and Claire Hines, *Fantasy* (London: Routledge, 2011).

84. C. W. Sullivan III, "High Fantasy," in *International Companion Encyclopedia of Children's Literature*, ed. Peter Hunt (London: Routledge, 2004): 436–446.

85. Helen Young, "Approaches to Medievalism: A Consideration of Taxonomy and Methodology through Fantasy Fiction," *Parergon* 27, no. 1 (2010): 163–179.

86. Kim Selling, "Fantastic Neomedievalism: The Image of the Middle Ages in Popular Fantasy," in *Flashes of the Fantastic: Selected Papers from "The War of the Worlds" Centennial, Nineteenth International Conference on the Fantastic in the Arts*, ed. David Ketterer (Westport: Praeger, 2004), 211–218.

87. William J. Sadler and Ekaterina V. Haskins, "Metonymy and the Metropolis: Television Show Settings and the Image of New York City," *Journal of Communication Inquiry* 29, no. 3 (2005): 195–216.

Chapter 2

1. McGinn, *Mindsight*, 13.

2. Rojek, "Indexing," 53.

3. Ibid.

4. Hennig, "Tourism," 185.

5. Reijnders, "Places of the Imagination," 14.

6. Ibid., 16.

7. McGinn, *Mindsight*, 49.

8. Hills, *Fan Cultures*, 137.

9. Ibid., 149.

10. Roesch, *The Experiences*, 114.

11. Ibid., 160.

12. Crouch, "Places Around Us," 67–72.

13. Garner, "Finding Nemo's Spaces," 20.

14. Lee, "Have Magic," 61.

15. Jones, "Theorizing Dissonance," 66.

16. Couldry, *Media Power*, 88–103.

17. Robert Moses Peaslee and Rosalynn Vasquez, "*Game of Thrones*, Tourism, and the Ethics of Adaptation," *Adaptation* (2020): apaa012, https://doi.org/10.1093/adaptation/apaa012.

18. Hills, *Fan Cultures*, 151.

19. Edensor, "Performing Tourism," 327.

20. Hills, *Fan Cultures*, 150.

21. Rebecca Williams, "Funko Hannibal in Florence: Fan Tourism, Participatory Culture, and Paratextual Play," *JOMEC Journal* 14 (2019): 82.

22. Crouch, "Places Around Us," 68.

23. Reijnders, *Places of the Imagination*, 81–104.

24. Buchmann, Moore, and Fisher, "Experiencing Film Tourism," 240.

25. Couldry, *Media Power*, 88–120.

26. Beeton, *Film-Induced Tourism*, 225–231.

27. Sangkyun Kim, "Audience Involvement and Film Tourism Experiences: Emotional Places, Emotional Experiences," *Tourism Management* 33 (2012): 387–396.

28. Jones, "Theorizing Dissonance," 58.

29. Stephen Joyce, "Media Tourism and Conflict Heritage in Dubrovnik, Westeros," *The Journal of Popular Culture* 52, no. 6 (2019): 1400.

30. Maria Sachiko Cecire, "Medievalism, Popular Culture and National Identity in Children's Fantasy Literature," *Studies in Ethnicity and Nationalism* 9, no. 5 (2009): 398.

31. Jenkins, *Textual Poachers*, 11.

32. Richard Butler, "It's Only Make Believe: The Implications of Fictional and Authentic Locations in Films," *Worldwide Hospitality and Tourism Themes* 3, no. 2 (2011): 91–101; Glen Croy, "Film Tourism," in *Special Interest Tourism: Concepts, Contexts and Cases*, ed. Sheila Agarwal, Graham Busby, and Ron Huang (Wallingford: CABI International, 2018), 85–96; Niki Macionis and Beverley Sparks, "Film-Induced Tourism: An Incidental Experience," *Tourism Review International* 13, no. 2 (2009): 93–101.

33. Croy, "Film Tourism: Sustained Economic Contributions," 159–164.

34. Christine Lundberg, Vassilios Ziakas, and Nigel Morgan, "Conceptualising On-Screen Tourism Destination Development," *Tourist Studies* 18, no. 1 (2018): 83–104.

35. Joyce, "Media Tourism and Conflict Heritage."

36. Ernest Mathijs and Martin Barker, "Seeing the Promised Land from Afar: The Perception of New Zealand by Overseas *The Lord of the Rings* Audiences," in *How We Became Middle-earth: A Collection of Essays on "The Lord of the Rings,"* ed. Adam Lam and Natalia Oryshchuk (Switzerland: Walking Tree Publishers, 2007), 107–128.

37. Buchmann, Moore, and Fisher, "Experiencing Film Tourism," 239.

38. Lee, "Have Magic."

39. Hills, *Fan Cultures*.

40. Will Brooker, "Everywhere and Nowhere: Vancouver, Fan Pilgrimage, and the Urban Imaginary," *International Journal of Cultural Studies* 10, no. 4 (2007): 423–444.

41. Hills, *Fan Cultures*, 148.

42. Garner, "Finding Nemo's Spaces," 23.

43. Joyce, "Media Tourism and Conflict Heritage."

44. K. J. Donnelly, "'Troubles Tourism': The Terrorism Theme Park On and Off Screen," in *The Media and the Tourist Imagination: Converging Cultures*, ed. David Crouch, Rhona Jackson, and Felix Thompson (Oxon: Routledge, 2005): 92–104.

45. Ipek A. Çelik Rappas and Stefano Baschiera, "Fabricating 'Cool' Heritage for Northern Ireland: *Game of Thrones* Tourism," *The Journal of Popular Culture* 53, no. 3 (2020): 648–666.

46. Anne Buchmann and Warwick Frost, "Wizards Everywhere? Film Tourism and the Imagining of National Identity in New Zealand," In *Tourism and National Identities: An International Perspective*, ed. Elspeth Frew and Leanne White (Oxon: Routledge, 2011), 52–64.

47. Hills, *Fan Cultures*, 148.

Chapter 3

1. Gundolf Graml, "(Re)Mapping the Nation: *Sound of Music* Tourism and National Identity in Austria, ca 2000 CE," *Tourist Studies* 4, no. 2 (2004): 137–159; Roesch, *The Experiences*.

2. Reijnders, *Places of the Imagination*, 55–80.

3. Brooker, "The *Blade Runner* Experience."

4. Sandvoss, *Fans*, 96.

5. Williams, *Post-Object Fandom*, 17–28.

6. C. Lee Harrington and Denise D. Bielby, "Autobiographical Reasoning in Long-Term Fandom," *Transformative Works and Cultures* 5 (2010), https://doi.org/10.3983/twc.2010.0209; C. Lee Harrington and Denise D. Bielby, "A Life Course Perspective on Fandom," *International Journal of Cultural Studies* 13, no. 5 (2010): 429–450.

7. C. Lee Harrington, Denise D. Bielby, and Anthony R. Bardo, "Life Course Transitions and the Future of Fandom," *International Journal of Cultural Studies* 14, no. 6 (2011): 567–590.

8. Rodman, *Elvis after Elvis*, 99.

9. Yi-Fu Tuan, *Space and Place: The Perspective of Experience* (Minneapolis: University of Minnesota Press, 1977), 182.

10. David Seamon, *A Geography of the Lifeworld: Movement, Rest, and Encounter* (London: Croom Helm, 1979), 57.

11. Tuan, *Space and Place*, 185.

12. See Setha M. Low and Irving Altman, "Place Attachment: A Conceptual Inquiry," in *Place Attachment*, ed. Irving Altman and Setha M. Low (New York: Plenum Press, 1992), 1–12, and Leila Scannell and Robert Gifford, "Defining Place Attachment: A Tripartite Organizing Framework," *Journal of Environmental Psychology* 30, no. 1 (2010): 1–10.

13. Sandvoss, *Fans*, 64.

14. See Pauline Garvey, "The Norwegian Country Cabin and Functionalism: A Tale of Two Modernities," *Social Anthropology* 16, no. 2 (2008): 203–220, and Keith Halfacree, "'A Solid Partner in a Fluid World' and/or 'Line of Flight'? Interpreting Second Homes in the Era of Mobilities," *Norsk Geografisk Tidsskrift–Norwegian Journal of Geography* 65, no. 3 (2011): 144–153.

15. Jenkins, *Textual Poachers*, 286–291.

16. Sue Short, *Cult Telefantasy Series: A Critical Analysis of "The Prisoner," "Twin Peaks," "The X-Files," "Buffy the Vampire Slayer," "Lost," "Heroes," "Doctor Who" and "Star Trek"* (Jefferson: McFarland, 2011).

17. Erin Hanna, "Be Selling You: *The Prisoner* as Cult and Commodity," *Television & New Media* 15, no. 5 (2014): 433–448.

18. Ibid., 439.

19. Hills, *Fan Cultures*, 134.

20. Clough Williams-Ellis, *Portmeirion: The Place and Its Meaning* rev. ed. (Porthmadog: Portmeirion Shops, 2014).

21. Roger Langley, *Portmeirion on Film and Television—The Full 70 Year Screen History* (available from the shops in Portmeirion, 2011).

22. Marc Augé, *Non-Places: An Introduction to Supermodernity*, trans. John Howe (London: Verso, 2008).

23. Rodman, *Elvis after Elvis*, 124.

24. Hills, *Fan Cultures*, 155.

25. Geraghty, *Cult Collectors*, 93–119.

26. Jennifer E. Porter, "Pilgrimage and the IDIC Ethic: Exploring *Star Trek* Convention Attendance as Pilgrimage," in *Intersecting Journeys: The Anthropology of Pilgrimage and Tourism*, ed. Ellen Badone and Sharon R. Roseman (Urbana: University of Illinois Press, 2004), 160–179.

27. Seamon, *A Geography*, 151.

28. Rodman, *Elvis after Elvis*.

29. Roger C. Aden, *Popular Stories and Promised Lands: Fan Cultures and Symbolic Pilgrimages* (Tuscaloosa: University of Alabama Press, 1999).

30. Sandvoss, "I♥Ibiza."

31. Victor Turner, *The Ritual Process: Structure and Anti-Structure* (Ithaca: Cornell University Press, 1977).

32. Rodman, *Elvis after Elvis*, 124.

33. Justine Digance, "Religious and Secular Pilgrimage: Journeys Redolent with Meaning," in *Tourism, Religion and Spiritual Journeys*, ed. Dallen J. Timothy and Daniel H. Olsen (London: Routledge, 2006), 40.

34. Sandvoss, *Fans*, 126.

35. Mittermeier, "(Un)Conventional Voyages?," 178.

36. Doreen Massey, *Place, Space, and Gender* (Cambridge: Polity, 1994), 155.

37. Williams, *Post-Object Fandom*, 20–22.

38. Ibid., 2.

39. Ibid., 26.

40. Abby Waysdorf, "'I Don't Think the Two Would Be the Same without Each Other': Portmeirion as Unintentional Paratext," *JOMEC Journal* 14 (2019): 33–52.

41. Tim Cresswell, *Place: An Introduction* (Chichester: Wiley Blackwell, 2014), 120.

42. Derek H. Alderman, Stefanie K. Benjamin, and Paige P. Schneider, "Transforming Mount Airy into Mayberry: Film-Induced Tourism as Place-Making," *Southeastern Geographer* 52, no. 2 (2012): 212–239.

Chapter 4

1. Niallgirlalmighty, Tumblr post, July 23, 2014, http://niallgirlalmighty.tumblr.com/post/9268173743 (accessed October 12, 2015).

2. Susan G. Davis, "The Theme Park: Global Industry and Cultural Form," *Media, Culture & Society* 18, no. 3 (1996): 399–422.

3. Meyrav Koren-kuik, "Desiring the Tangible: Disneyland, Fandom, and Spatial Immersion," in *Fan CULTure: Essays on Participatory Fandom in the 21st Century*, ed. Kristin M. Barton and Jonathan Malcolm Lampley (Jefferson: McFarland, 2014), 146–158.

4. Rebecca Williams, *Theme Park Fandom: Spatial Transmedia, Materiality and Participatory Cultures* (Amsterdam: Amsterdam University Press, 2020), 12.

5. Saler, *As-If*, 14.

6. Umberto Eco, *Travels in Hyperreality: Essays*, trans. William Weaver (San Diego: Harcourt Brace, 1986), 44.

7. Michael Sorkin, "See You in Disneyland," in *Variations on a Theme Park: The New American City and the End of Public Space*, ed. Michael Sorkin (New York: Hill and Wang, 1992), 216.

8. Aden, *Popular Stories*, 228.

9. Ibid., 234.

10. Hills, *Fan Cultures*, 151.

11. Rachel Marie Gilbert, "A Potterhead's Progress: A Quest for Authenticity at the Wizarding World of Harry Potter," in *Playing Harry Potter: Essays and Interviews on Fandom and Performance*, ed. Lisa S. Brenner (Jefferson: McFarland, 2015), 35.

12. Scott A. Lukas, "Theming as a Sensory Phenomenon: Discovering the Senses on the Las Vegas Strip," in *The Themed Space: Locating Culture, Nation, and Self*, ed. Scott Lukas (Plymouth: Lexington, 2007), 82.

13. Salvador Anton Clavé, *The Global Theme Park Industry* (Wallingford: CABI, 2007), 178.

14. Ibid.

15. Ibid., 193.

16. Ryan, *Narrative*, 41.

17. Williams, *Theme Park Fandom.*

18. Williams, *Theme Park Fandom*, 12.

19. Davis, "The Theme Park."

20. Miodrag Mitrasinovic, *Total Landscape, Theme Parks, and Public Space* (Aldershot: Ashgate, 2006), 42–110.

21. Koren-kuik, "Desiring the Tangible," 152–154.

22. Mitrasinovic, *Total Landscape*, 42–110.

23. Beeton, *Film-Induced Tourism*, 237–264.

24. Williams, *Theme Park Fandom*, 101–131.

25. Erkki Huhtamo, "Encapsulated Bodies in Motion: Simulators and the Quest for Total Immersion," in *Critical Issues in Electronic Media*, edited by Simon Penny (Albany: State University of New York Press, 1995), 160.

26. Constance Balides, "Immersion in the Virtual Ornament: Contemporary 'Movie Ride' Films," in *Rethinking Media Change: The Aesthetics of Transition*, ed. David Thorburn and Henry Jenkins (Cambridge, MA: MIT Press, 2003), 315–336; Huhtamo, "Encapsulated Bodies."

27. Ryan, *Narrative*, 41.

28. Ibid.

29. Saler, *As-If*, 7.

30. Williams, *Theme Park Fandom*, 187.

31. See Lee, "Have Magic."

32. Keidl, "Behind-the-(Museum)-Scenes"; Mortensen, "Commemorating Popular Media Heritage."

33. Matthew Freeman, "Transmedia Attractions: The Case of Warner Bros Studio Tour—The Making of Harry Potter," in *The Routledge Companion to Transmedia Studies*, ed. Matthew Freeman and Renira Rampazzo Gambarato (New York: Routledge, 2018), 124–130.

34. Susan Gunelius, *Harry Potter: The Story of a Global Business Phenomenon* (Houndmills: Palgrave Macmillan, 2008), 36–37.

35. Mitrasinovic, *Total Landscape*, 42–110.

36. Freeman, "Transmedia Attractions."

37. Katherine Larsen, "(Re)Claiming Harry Potter Fan Pilgrimage Sites," in *Playing Harry Potter: Essays and Interviews on Fandom and Performance*, ed. Lisa S. Brenner (Jefferson: McFarland, 2015), 46.

38. Larsen, "(Re)Claiming Harry Potter," 45–52.

39. Reijnders, *Places of the Imagination*, 14.

40. Crouch, "Places Around Us," 68.

41. Lukas, "Theming as a Sensory Phenomenon."

42. Williams, *Theme Park Fandom*.

43. Saler, *As-If*, 30.

44. Norman Klein, "Electronic Baroque: Jerde Cities," in *You Are Here: The Jerde Partnership International*, ed. Ray Bradbury (London: Phaidon, 1999), http://artefact.mi2.hr/_a04/lang_en/theory_klein_en.htm.

45. Aden, *Popular Stories*, 234.

46. Klein, "Electronic Baroque."

47. Janet H. Murray, *Hamlet on the Holodeck: The Future of Narrative in Cyberspace* (New York: The Free Press, 1997); Ryan, *Narrative*.

48. Jason Mittell, "Strategies of Storytelling on Transmedia Television," in *Storyworlds Across Media: Towards a Media-Conscious Narratology*, ed. Marie-Laure Ryan and Jan-Noël Thon (Lincoln: University of Nebraska Press, 2014), 259.

49. Williams, *Theme Park Fandom*, 163.

50. Turner, *The Ritual Process*.

51. See Hills, *Fan Cultures*, and Brooker, "Everywhere and Nowhere."

52. Saler, *As-If*, 55.

53. Jenkins, *Textual Poachers*.

54. See also Williams, *Theme Park Fandom*, 118–121.

55. Christian Sylt, "Revealed: The World's Fastest-Growing Theme Park," *Forbes*, June 30, 2019, https://www.forbes.com/sites/csylt/2019/06/30/re vealed-the-worlds-fastest-growing-theme-park/.

Chapter 5

1. Allegra Frank, "*Friends* Is Leaving Netflix in 2020," *Vox*, July 9, 2019, https://www.vox.com/culture/2019/7/9/20687923/friends-leaving-netflix -2020-streaming-hbo-max; Simone Knox and Kai Hanno Schwind, *Friends: A Reading of the Sitcom* (Cham: Palgrave Macmillan, 2019), 2.

2. Sean Coughlan, "The One about Friends Still Being Most Popular," *BBC News*, January 30, 2019, https://www.bbc.com/news/education-47043831; Andrew Dalton, "25 Years Later, a New Generation Gets Immersed in 'Friends,'" *Associated Press*, September 22, 2019, https://apnews.com/42cf0d6 a9c3d42bf89e28a7a6863932f.

3. See Mark Jancovich, "Cult Fictions: Cult Movies, Subcultural Capital, and the Production of Cultural Distinctions," *Cultural Studies* 16, no. 2 (2002): 306–322.

4. Linden and Linden, *Fans and Fan Cultures*, 38–40.

5. Jenkins, *Textual Poachers*.

6. Karen Hellekson, "A Fannish Field of Value: Online Fan Gift Culture," *Cinema Journal* 48, no. 4 (2009): 113–118; Francesca Coppa, "Fuck Yeah, Fandom Is Beautiful," *The Journal of Fandom Studies* 2, no. 1 (2014): 73–82.

7. Jon Sundbo and Flemming Sørenson, "Introduction to the Experience Economy," in *Handbook on the Experience Economy*, ed. Jon Sundbo and Flemming Sørenson (Cheltenham: Edward Elgar, 2013): 1–17; Linden and Linden, *Fans and Fan Cultures*, 47–50.

8. Matt Hills, "Torchwood's Trans-transmedia: Media Tie-ins and Brand 'Fanagement,'" *Participations* 9, no. 2 (2012): 409–428.

9. Joli Jenson, "Fandom as Pathology: The Consequences of Characterization," in *The Adoring Audience: Fan Culture and Popular Media*, ed. Lisa A. Lewis (London: Routledge, 1992), 13.

10. Roberta Pearson, "Bachies, Bardies, Trekkies, and Sherlockians," in *Fandom: Identities and Communities in a Mediated World*, ed. C. Lee Harrington, Jonathan Gray, and Cornel Sandvoss (New York: New York University Press, 2007), 98–109.

11. Alan McKee, "The Fans of Cultural Theory," in *Fandom: Identities and Communities in a Mediated World*, ed. C. Lee Harrington, Jonathan Gray, and Cornel Sandvoss (New York: New York University Press, 2007), 88–97.

12. Hellekson, "A Fannish Field."

13. Geraghty, *Cult Collectors*, 30.

14. Stanfill, *Exploiting Fandom*, 3–11.

15. Booth, *Playing Fans*, 1.

16. See Henry Jenkins, "The Future of Fandom," in *Fandom: Identities and Communities in a Mediated World*, ed. C. Lee Harrington, Jonathan Gray, and Cornel Sandvoss (New York: New York University Press, 2007), 357–364.

17. Stanfill, *Exploiting Fandom*, 5.

18. Ibid.

19. Sara Gwenllian-Jones, "Web Wars: Resistance, Online Fandom, and Studio Censorship," in *Quality Popular Television: Cult TV, the Industry and Fans*, ed. Mark Jancovich and James Lyons (London: British Film Institute, 2003), 165.

20. Linden and Linden, *Fans and Fan Cultures*, 76.

21. Ibid., 4.

22. Paul Booth, "Framing Alterity: Reclaiming Fandom's Marginality," in "The Future of Fandom," special 10th anniversary issue, *Transformative Works and Cultures* 28 (2018), https://doi.org/10.3983/twc.2018.1420.

23. Judy Kutulas, "Anatomy of a Hit: *Friends* and Its Sitcom Legacies," *The Journal of Popular Culture* 51, no. 5 (2018): 1172–1189.

24. Knox and Schwind, *Friends*, 7–11.

25. Anne Marie Todd, "Saying Goodbye to *Friends*: Situation Comedy as Lived Experience," *The Journal of Popular Culture* 44, no. 4 (2011): 855–872.

26. Kutulas, "Anatomy of a Hit," 1172–1189.

27. Adam Sternbergh, "Is *Friends* Still the Most Popular Show on TV?," *Vulture*, March 21, 2016, https://www.vulture.com/2016/03/20-somethings -streaming-friends-c-v-r.html.

28. Coughlan, "The One"; Dalton, "25 Years Later."

29. Danielle Turchiano, "'Friends' at 25: How Warner Bros. TV Built an Experiential Empire," *Variety*, September 19, 2019, https://variety.com/2019 /tv/features/friends-25th-anniversary-stunts-partnerships-popup-lego-apps -pottery-barn-studio-tour-interview-1203331201/.

30. Catherine Johnson, "Tele♥branding in TVIII: The Network as Brand and the Programme as Brand," *New Review of Film and Television Studies* 5, no. 1 (2007): 7–8.

31. Matt Hills, *"Doctor Who": The Unfolding Event—Marketing, Merchandising and Mediatizing a Brand Anniversary* (Basingstoke: Palgrave Macmillan, 2015), 4.

32. Geraghty, *Cult Collectors*; Lincoln Geraghty, "Nostalgia, Fandom, and the Remediation of Children's Culture," in *A Companion to Media Fandom and Fan Studies*, ed. Paul Booth (Hoboken: Wiley Blackwell, 2018), 161–174.

33. Hills, *The Unfolding Event*, 4.

34. Joseph B. Pine and James H. Gilmore, *The Experience Economy: Work Is Theatre & Every Business a Stage* (Boston: Harvard Business Press, 1999); Sundbo and Sørensen, "Introduction."

35. Sundbo and Sørensen, "Introduction," 7.

36. Pine and Gilmore, *The Experience Economy*, 20–25.

37. Booth, *Playing Fans*, 101–122.

38. Ibid., 23.

39. Catherine Tosenberger, "Homosexuality at the Online Hogwarts: Harry Potter Slash Fanfiction," *Children's Literature* 36, no. 1 (2008): 185–207.

40. Lucy Bennett, "Representations of Fans and Fandom in the British Newspaper Media," in *A Companion to Media Fandom and Fan Studies*, ed. Paul Booth (Hoboken: Wiley Blackwell, 2018), 107.

41. Jonathan Gray, *Show Sold Separately: Promos, Spoilers, and Other Media Paratexts* (New York: New York University Press, 2010), 47–80.

42. Joan Leach, "Rhetorical Analysis," in *Qualitative Researching with Text, Image and Sound: A Practical Handbook*, ed. Martin W. Bauer and George Gaskell (London: Sage, 2000), 207–226; Lisa Ellen Silvestri, "Memeingful Memories and the Art of Resistance," *New Media & Society* 20, no. 11 (2018): 3997–4016; Mark Zachry, "Rhetorical Analysis," in *The Handbook of Business Discourse*, ed. Francesca Bargiela-Chiappini (Edinburgh: Edinburgh University Press, 2009), 68–79.

43. Danielle O'Brien, "COULD WE BE MORE EXCITED? FriendsFest Is Returning to the UK . . . Here's What to Expect and How You Can Get Tickets NOW," *The Sun*, March 28, 2018, https://www.thesun.co.uk/fabulous /5918782/friendsfest-is-returning-to-the-uk-heres-what-to-expect-and-how -you-can-get-tickets/.

44. Clarisse Loughrey, "Friends: FriendsFest Is Back and Touring the UK This Time," *The Independent*, June 28, 2016, https://www.independent.co.uk /arts-entertainment/tv/news/friends-friendsfest-is-back-and-touring-the-uk -this-time-a7106981.html.

45. Rhiannon Evans, "FriendsFest: The One Where Fans Wander around Monica's Apartment," *The Guardian*, September 16, 2015, https://www.the guardian.com/tv-and-radio/shortcuts/2015/sep/16/friendsfest-exhibition -one-where-fans-wander-monicas-apartment.

46. Bruce Haring, "'Friends' 25th Anniversary Spawns a Pop-Up Experience Straight Outta 1994," *Deadline*, July 29, 2019, https://deadline.com/2019/07/ friends-pop-up-experience-tickets-1202655212/.

47. Jackie Strause, "'Friends' Turns 25: Inside the NBC Sitcom's Pop-Up Return to New York," *The Hollywood Reporter*, September 6, 2019, https://www .hollywoodreporter.com/live-feed/friends-turns-25-inside-new-york-city-pop-up -1237174.

48. Brooke Lefferts, "'Friends' Pop-Up Lets Sitcom's Fans Explore Show's Key Props," *Associated Press*, September 6, 2019, https://apnews.com/ceaf 5e179abe494894eb3fe90551eb5e.

49. Lauren Sarner, "Overhyped 'Friends' Pop-Up Is a 25th Anniversary Rip-Off," *New York Post*, September 6, 2019, https://nypost.com/2019/09/06/over hyped-friends-pop-up-is-a-25th-anniversary-rip-off/.

50. Kyle Turner, "The *Friends* Pop-Up Captures Everything About the Show Except What Made It Great," *Slate*, September 13, 2019, https://slate.com/cul ture/2019/09/friends-tv-pop-up-experience-shop-new-york.html.

51. Quoted in Madeleine Buckley, "Step into the World of 'Friends' at the NYC Pop-Up," *The Pop Insider*, September 7, 2019, https://thepopinsider.com /news/first-look/friends-pop-up-review/.

52. Amy Duncan, "The One Where a Friends-Obsessed Couple Got Engaged in Monica and Chandler's Apartment at FriendsFest," *MetroUK*, August 30, 2016, https://metro.co.uk/2016/08/30/the-one-where-a-friends-obsessed-cou ple-got-engaged-in-monica-and-chandlers-apartment-at-friendsfest-6099408/.

53. Damien Lucas, "Legendary TV Show Friends Takes Over Milton Keynes This Week for 25th Anniversary Summer Tour," *Milton Keynes Citizen*, September 4, 2019, https://www.miltonkeynes.co.uk/whats-on/things-to-do /legendary-tv-show-friends-takes-over-milton-keynes-week-25th-anniversary -summer-tour-944459.

54. Catriona Harvey-Jenner and Dusty Baxter-Wright, "Hooray, FRIENDS-FEST Is Back for a 12 Week Tour of the UK," *Cosmopolitan*, July 20, 2017, https://www.cosmopolitan.com/uk/entertainment/news/a44301/friends-fest -touring-round-country/.

55. Elizabeth Adetula, "The One Where You Get to Have a Tour of Joey and

Chandler's Apartment This Time . . . Yep, FriendsFest Is Back," *MetroUK*, February 21, 2017, https://metro.co.uk/2017/02/21/the-one-where-you-get-to-have-a-tour-of-joey-and-chandlers-apartment-this-time-yep-friendsfest-is-back-6462736/.

56. Madison Flager, "Take a Look Inside the 'Friends' Pop-Up Now Open in New York City," *Delish*, September 17, 2019, https://www.delish.com/food/a28622614/friends-pop-up-new-york-city-warner-bros/.

57. Jen Abidor, Nora Dominick, and Lauren Yapalater, "We Went to the 'Friends' Pop-Up and Could It BE Any More Perfect?!," *BuzzFeed*, September 18, 2019, https://www.buzzfeed.com/jenniferabidor/heres-everything-you-need-to-know-about-the-friends-nyc-pop.

58. Dianne Bourne and Katie Fitzpatrick, "Soap Stars Flock to the FriendsFest at Heaton Park," *Manchester Evening News*, August 9, 2017, https://www.manchestereveningnews.co.uk/news/showbiz-news/coronation-street-emmerdale-friendsfest-friends-13452667; Julia Pritchard, "Oh, My, God! Lydia Bright Flashes Some Leg in Floral Maxi as She Fangirls over Childhood Favourite Maggie Wheeler (aka Janice) at Friendsfest," *MailOnline*, August 24, 2016, https://www.dailymail.co.uk/tvshowbiz/article-3755619/Lydia-Bright-flashes-leg-floral-maxi-fangirls-childhood-favourite-Maggie-Wheeler-aka-Janice-Friendsfest.html.

59. Knox and Schwind, *Friends*, 125–163.

60. Ibid.

61. Lauren Jade Thompson, "'It's Like a Guy Never Lived Here!': Reading the Gendered Domestic Spaces of *Friends*," *Television & New Media* 19, no. 8 (2018): 758–774.

62. Lefferts, "'Friends' Pop-Up."

63. Henry Jenkins, *Convergence Culture: Where Old and New Media Collide* (New York: New York University Press, 2006), 95–96.

64. Stanfill, *Exploiting Fandom*, 91.

65. Ibid., 95.

66. Emily Heward, "FriendsFest Is Coming to Manchester This Weekend—with More Room Sets and THAT Pivot Scene," *Manchester Evening News*, July 30, 2018, https://www.manchestereveningnews.co.uk/whats-on/whats-on-news/friends-fest-manchester-tickets-heaton-14968617.

67. Lucas, "Legendary TV Show."

68. Heward, "FriendsFest Is Coming."

69. Buckley, "Step into the World."

70. *People*, "Explore the Instagram Worthy, Sold-Out *Friends* 25th Anniversary Pop-Up!" September 19, 2019, https://people.com/tv/explore-the-instagram-worthy-sold-out-friends-25th-anniversary-pop-up/.

71. Jansson, "Spatial Phantasmagoria"; Linden and Linden, *Fans and Fan Cultures*, 115–121.

72. Stanfill, *Exploiting Fandom*, 97.

73. Linden and Linden, *Fans and Fan Cultures*, 50.

74. Stanfill, *Exploiting Fandom*, 99.

75. Ibid., 102.

76. *The Argus*, "Friendsfest—Everything You Need to Know," September 7, 2018, https://www.theargus.co.uk/news/16695677.friendsfest-starts-at-preston-park-today-all-you-need-to-know/.

77. Abidor, Dominick, and Yapalater, "We Went to the 'Friends' Pop-Up."

78. Catherine Gee, "Friends Fans: FriendsFest Has Arrived—What to Expect and How to Get Tickets," *The Telegraph*, August 24, 2016, https://www.telegraph.co.uk/tv/2016/06/28/friendsfest-is-back-and-its-going-on-tour--how-to-get-to-tickets/.

79. Kristina Busse, *Framing Fan Fiction: Literary and Social Practices in Fan Fiction Communities* (Iowa City: University of Iowa Press, 2017), 177–196; Mel Stanfill, "'They're Losers, but I Know Better': Intra-Fandom Stereotyping and the Normalization of the Fan Subject," *Critical Studies in Media Communication* 30, no. 2 (2013): 117–134.

80. Ryan Lizardi, *Mediated Nostalgia: Individual Memory and Contemporary Mass Media* (Lanham: Lexington Books, 2015).

81. Buckley, "Step into the World."

82. Geraghty, "Nostalgia, Fandom," 171.

83. Mortensen, "Commemorating Popular Media Heritage."

84. Busse, *Framing Fan Fiction*, 8–9.

Conclusion

1. Urry and Larsen, *The Tourist Gaze 3.0*, 5.

2. Adler, "Origins of Sightseeing."

3. Urry and Larsen, *The Tourist Gaze 3.0*, 21.

4. Grossberg, "Is There a Fan?"; Hills, *Fan Cultures*, 105.

5. Roesch, *The Experiences*, 114.

6. Hills, *Fan Cultures*, 54.

7. Couldry, *Media Power*, 88–120.

8. Beeton, *Film-Induced Tourism*, 237–264.

9. Hills, *Fan Cultures*.

10. Alderman, Benjamin, and Schneider, "Transforming Mount Airy into Mayberry"; Joyce, "Media Tourism and Conflict Heritage"; Tom Mordue, "Performing and Directing Resident/Tourist Cultures in Heartbeat Country," *Tourist Studies* 1, no. 3 (2001): 233–252.

11. Williams, *Theme Park Fandom*.

12. See Chris Agar, "Disney Addresses Lower Attendance at Star Wars

Galaxy's Edge," *Screenrant*, August 7, 2019, https://screenrant.com/star-wars-galaxys-edge-attendance-low-disney/; Todd Martens, "Commentary: What Works, What's Missing and What Needs Fixing at Disney's Galaxy's Edge," *Los Angeles Times*, October 7, 2019, https://www.latimes.com/entertainment-arts/story/2019-10-07/disneyland-star-wars-galaxys-edge-progress-report; Lilliana Parker, "Editorial: Why Star Tours Is a Better Star Wars Attraction Than Millennium Falcon," *MiceChat*, December 19, 2019, https://www.micechat.com/238932-editorial-why-star-tours-is-a-better-star-wars-attraction-than-millennium-falcon/; James Whitbrook, "*Star Wars: Galaxy's Edge* Is Almost Too Alien for Its Own Good," *io9*, November 27, 2019, https://io9.gizmodo.com/star-wars-galaxys-edge-is-almost-too-alien-for-its-own-1840053738.

13. Rebecca Williams, "Theme Parks in the Time of the COVID-19 Pandemic," in *Pandemic Media: Preliminary Notes toward an Inventory*, ed. Philipp Dominik Keidl et al. (Meson Press, 2020), https://pandemicmedia.meson.press/chapters/space-scale/theme-parks-in-the-time-of-the-covid-19-pandemic/.

14. Jenkins, *Textual Poachers*.

15. Jenkins, *Convergence Culture*.

16. Booth, *Playing Fans*.

17. Stanfill, *Exploiting Fandom*.

Bibliography

Abidor, Jen, Nora Dominick, and Lauren Yapalater. "We Went to the 'Friends' Pop-Up and Could It BE Any More Perfect?!" *BuzzFeed*, September 18, 2019. https://www.buzzfeed.com/jenniferabidor/heres-everything-you-need-to-know-about-the-friends-nyc-pop.

Aden, Roger C. *Popular Stories and Promised Lands: Fan Cultures and Symbolic Pilgrimages*. Tuscaloosa: University of Alabama Press, 1999.

Adetula, Elizabeth. "The One Where You Get to Have a Tour of Joey and Chandler's Apartment This Time . . . Yep, FriendsFest Is Back." *MetroUK*, February 21, 2017. https://metro.co.uk/2017/02/21/the-one-where-you-get-to-have-a-tour-of-joey-and-chandlers-apartment-this-time-yep-friendsfest-is-back-6462736/.

Adler, Judith. "Origins of Sightseeing." *Annals of Tourism Research* 16, no. 1 (1989): 7–29.

Agar, Chris. "Disney Addresses Lower Attendance at Star Wars Galaxy's Edge." *Screenrant*, August 7, 2019. https://screenrant.com/star-wars-galaxys-edge-attendance-low-disney/.

Ah-young, Chung. "'Descendants of the Sun' Makes Taekbaek a Tourism Magnet." *The Korea Times*, March 24, 2016. https://www.koreatimes.co.kr/www/news/nation/2016/03/116_201065.html.

Alderman, Derek H., Stefanie K. Benjamin, and Paige P. Schneider. "Transforming Mount Airy into Mayberry: Film-Induced Tourism as Place-Making." *Southeastern Geographer* 52, no. 2 (2012): 212–239.

The Argus. "Friendsfest—Everything You Need to Know." September 7, 2018. https://www.theargus.co.uk/news/16695677.friendsfest-starts-at-preston-park-today-all-you-need-to-know/.

Attebery, Brian. *Strategies of Fantasy*. Bloomington: Indiana University Press, 1992.

Augé, Marc. *Non-Places: An Introduction to Supermodernity*. Translated by John Howe. London: Verso, 2008.

Auslander, Philip. *Liveness: Performance in a Mediatized Culture*. London: Routledge, 1999.

Badone, Ellen, and Sharon R. Roseman, eds. *Intersecting Journeys: The Anthropology of Pilgrimage and Tourism*. Urbana: University of Illinois Press, 2004.

Baker, Sarah, Lauren Istvandity, and Raphaël Nowak. "The Sound of Music Heritage: Curating Popular Music in Music Museums and Exhibitions." *International Journal of Heritage Studies* 22, no. 1 (2015): 70–81.

Balides, Constance. "Immersion in the Virtual Ornament: Contemporary 'Movie Ride' Films." In *Rethinking Media Change: The Aesthetics of Transition*, edited by David Thorburn and Henry Jenkins, 315–336. Cambridge, MA: MIT Press, 2003.

Beattie, Melissa. "The '*Doctor Who* Experience' (2012–) and the Commodification of Cardiff Bay." In *New Dimensions of Doctor Who: Adventures in Space, Time, and Television*, edited by Matt Hills, 177–191. London: I.B. Tauris, 2013.

Beeton, Sue. "The Advance of Film Tourism." *Tourism and Hospitality Planning & Development* 7, no. 1 (2010): 1–6.

———. *Film-Induced Tourism*. 2nd ed. Bristol: Channel View Publications, 2016.

Bennett, Lucy. "Representations of Fans and Fandom in the British Newspaper Media." In *A Companion to Media Fandom and Fan Studies*, edited by Paul Booth, 107–122. Hoboken: Wiley Blackwell, 2018.

Bolderman, Leonieke. *Contemporary Music Tourism: A Theory of Musical Topophilia*. London: Routledge, 2020.

Bolter, Jay David, and Richard Grusin. *Remediation: Understanding New Media*. Cambridge, MA: MIT Press, 1999.

Boorstin, Daniel J. *The Image: A Guide to Pseudo-Events in America*. New York: Vintage Books, 1962.

Booth, Paul. "Framing Alterity: Reclaiming Fandom's Marginality." In "The Future of Fandom," special 10th anniversary issue, *Transformative Works and Cultures* 28 (2018). https://doi.org/10.3983/twc.2018.1420.

———. *Playing Fans: Negotiating Fandom and Media in the Digital Age*. Iowa City: University of Iowa Press, 2015.

Bourne, Dianne, and Katie Fitzpatrick. "Soap Stars Flock to the FriendsFest at Heaton Park." *Manchester Evening News*, August 9, 2017. https://www.man chestereveningnews.co.uk/news/showbiz-news/coronation-street-emmer dale-friendsfest-friends-13452667.

Bramhill, Nick. "Force Awakens at Skellig Michael as Busiest Ever Tourism Season Begins." *Irish Independent*, May 8, 2017. https://www.independent.ie

/irish-news/force-awakens-at-skellig-michael-as-busiest-ever-tourism-season
-begins-35690097.html.

Brooker, Will. "The *Blade Runner* Experience: Pilgrimage and Liminal Space." In *The Blade Runner Experience: The Legacy of a Science-Fiction Classic*, edited by Will Brooker, 47–107. London: Wallflower Press, 2005.

———. "Everywhere and Nowhere: Vancouver, Fan Pilgrimage, and the Urban Imaginary." *International Journal of Cultural Studies* 10, no. 4 (2007): 423–444.

Buchmann, Anne, and Warwick Frost. "Wizards Everywhere? Film Tourism and the Imagining of National Identity in New Zealand." In *Tourism and National Identities: An International Perspective*, edited by Elspeth Frew and Leanne White, 52–64. Oxon: Routledge, 2001.

Buchmann, Anne, Kevin Moore, and David Fisher. "Experiencing Film Tourism: Authenticity & Fellowship." *Annals of Tourism Research* 37, no. 1 (2010): 229–248.

Buckley, Madeleine. "Step into the World of 'Friends' at the NYC Pop-Up." *The Pop Insider*, September 7, 2019. https://thepopinsider.com/news/first-look/friends-pop-up-review/.

Busse, Kristina. *Framing Fan Fiction: Literary and Social Practices in Fan Fiction Communities*. Iowa City: University of Iowa Press, 2017.

Busse, Kristina, and Jonathan Gray. "Fan Cultures and Fan Communities." In *The Handbook of Media Audiences*, edited by Virginia Nightingale, 425–443. Chichester: Wiley Blackwell, 2011.

Butler, Richard. "It's Only Make Believe: The Implications of Fictional and Authentic Locations in Films." *Worldwide Hospitality and Tourism Themes* 3, no. 2 (2011): 91–101.

Carl, Daniela, Sara Kindon, and Karen Smith. "Tourists' Experiences of Film Locations: New Zealand as 'Middle-Earth.'" *Tourism Geographies* 9, no. 1 (2007): 49–63.

Cecire, Maria Sachiko. "Medievalism, Popular Culture and National Identity in Children's Fantasy Literature." *Studies in Ethnicity and Nationalism* 9, no. 5 (2009): 395–409.

Çelik Rappas, Ipek A., and Stefano Baschiera. "Fabricating 'Cool' Heritage for Northern Ireland: *Game of Thrones* Tourism." *The Journal of Popular Culture* 53, no. 3 (2020): 648–666.

Clavé, Salvador Anton. *The Global Theme Park Industry*. Wallingford: CABI, 2007.

Cohen, Erik. "Authenticity and Commoditization in Tourism." *Annals of Tourism Research* 15, no. 3 (1988): 371–386.

Connell, Joanne. "Film Tourism—Evolution, Progress, and Prospects." *Tourism Management* 33, no. 5 (2012): 1007–1029.

Connell, Joanne, and Denny Meyer. "Balamory Revisited: An Evaluation of the

Screen Tourism Destination-Tourist Nexus." *Tourism Management* 30, no. 2 (2009): 194–207.

Coppa, Francesca. "Fuck Yeah, Fandom Is Beautiful." *The Journal of Fandom Studies* 2, no. 1 (2014): 73–82.

Coughlan, Sean. "The One about Friends Still Being Most Popular." *BBC News*, January 30, 2019. https://www.bbc.com/news/education-47043831.

Couldry, Nick. *The Place of Media Power: Pilgrims and Witnesses of the Media Age.* London: Routledge, 2000.

Crang, Mike. "Knowing, Tourism and Practices of Vision." In *Leisure/Tourism Geographies: Practices and Geographical Knowledge*, edited by David Crouch, 238–257. London: Routledge, 1999.

Cresswell, Tim. *Place: An Introduction.* Chichester: Wiley Blackwell, 2014.

Crouch, David. "Places Around Us: Embodied Lay Geographies in Leisure and Tourism." *Leisure Studies* 19, no. 2 (2000): 63–76.

Croy, Glen W. "Film Tourism." In *Special Interest Tourism: Concepts, Contexts and Cases*, edited by Sheila Agarwal, Graham Busby, and Ron Huang, 85–96. Wallingford: CABI International, 2018.

———. "Film Tourism: Sustained Economic Contributions to Destinations." *Worldwide Hospitality and Tourism Themes* 3, no. 2 (2011): 159–164.

Dalton, Andrew. "25 Years Later, a New Generation Gets Immersed in 'Friends.'" *Associated Press*, September 22, 2019. https://apnews.com/42cf0d6a9c3d 42bf89e28a7a6863932f.

Davis, Susan G. "The Theme Park: Global Industry and Cultural Form." *Media, Culture & Society* 18, no. 3 (1996): 399–422.

Di Cesare, Francesco, Luca D'Angelo, and Gloria Rech. "Films and Tourism: Understanding the Nature and Intensity of Their Cause–Effect Relationship." *Tourism Review International* 13, no. 2 (2009): 103–111.

Digance, Justine. "Religious and Secular Pilgrimage: Journeys Redolent with Meaning." In *Tourism, Religion and Spiritual Journeys*, edited by Dallen J. Timothy and Daniel H. Olsen, 36–48. London: Routledge, 2006.

Donnelly, K. J. "'Troubles Tourism': The Terrorism Theme Park On and Off Screen." In *The Media and the Tourist Imagination: Converging Cultures*, edited by David Crouch, Rhona Jackson, and Felix Thompson, 92–104. Oxon: Routledge, 2005.

Duncan, Amy. "The One Where a Friends-Obsessed Couple Got Engaged in Monica and Chandler's Apartment at FriendsFest." *MetroUK*, August 30, 2016. https://metro.co.uk/2016/08/30/the-one-where-a-friends-obsessed -couple-got-engaged-in-monica-and-chandlers-apartment-at-friendsfest -6099408/.

Eco, Umberto. *Travels in Hyperreality: Essays.* Translated by William Weaver. San Diego: Harcourt Brace, 1986.

Edensor, Tom. "Performing Tourism, Staging Tourism: (Re)Producing Tourist Space and Practice." *Tourist Studies* 1, no. 1 (2001): 59–81.

Evans, Rhiannon. "FriendsFest: The One Where Fans Wander around Monica's Apartment." *The Guardian*, September 16, 2015. https://www.theguardian .com/tv-and-radio/shortcuts/2015/sep/16/friendsfest-exhibition-one-where -fans-wander-monicas-apartment.

Fernandez-Young, Anita, and Robert Young. "Measuring the Effects of Film and Television on Tourism to Screen Locations: A Theoretical and Empirical Perspective." *Journal of Travel & Tourism Marketing* 24, nos. 2–3 (2008): 195–212.

Fiske, John. "The Cultural Economy of Fandom." In *The Adoring Audience: Fan Culture and Popular Media*, edited by Lisa A. Lewis, 30–49. London: Routledge, 1992.

Flager, Madison. "Take a Look Inside the 'Friends' Pop-Up Now Open in New York City." *Delish*, September 17, 2019. https://www.delish.com/food/a28 622614/friends-pop-up-new-york-city-warner-bros/.

Fowkes, Katherine A. *The Fantasy Film*. Chichester: Wiley-Blackwell, 2010.

Frank, Allegra. "*Friends* Is Leaving Netflix in 2020." *Vox*, July 9, 2019. https:// www.vox.com/culture/2019/7/9/20687923/friends-leaving-netflix-2020 -streaming-hbo-max.

Freeman, Matthew. "Transmedia Attractions: The Case of Warner Bros Studio Tour—The Making of Harry Potter." In *The Routledge Companion to Transmedia Studies*, edited by Matthew Freeman and Renira Rampazzo Gambarato, 124–130. New York: Routledge, 2018.

Furby, Jacqueline, and Claire Hines. *Fantasy*. London: Routledge, 2011.

Garner, Ross. "Finding Nemo's Spaces: Defining and Exploring Transmedia Tourism.'" *JOMEC Journal* 14 (2019): 11–32.

———. "Symbolic and Cued Immersion: Paratextual Framing Strategies on the *Doctor Who* Experience Walking Tour." *Popular Communication* 14, no. 2 (2016): 86–98.

Garvey, Pauline. "The Norwegian Country Cabin and Functionalism: A Tale of Two Modernities." *Social Anthropology* 16, no. 2 (2008): 203–220.

Gee, Catherine. "Friends Fans: FriendsFest Has Arrived—What to Expect and How to Get Tickets." *The Telegraph*, August 24, 2016. https://www.telegraph .co.uk/tv/2016/06/28/friendsfest-is-back-and-its-going-on-tour--how-to -get-to-tickets/.

Geraghty, Lincoln. *Cult Collectors: Nostalgia, Fandom, and Collecting Popular Culture*. London: Routledge, 2014.

———. "Nostalgia, Fandom, and the Remediation of Children's Culture." *In A Companion to Media Fandom and Fan Studies*, edited by Paul Booth, 161–174. Hoboken: Wiley Blackwell, 2018.

Geraghty, Lincoln, Vassilios Ziakas, and Christine Lundberg. "Guest Editorial:

Exploring the Popular Culture and Tourism Place Making Nexus." *The Journal of Popular Culture* 52, no. 6 (2019): 1241–1249.

Giddens, Anthony. *Modernity and Self-Identity: Self and Society in the Late Modern Age*. Cambridge: Polity Press, 1991.

Gilbert, Rachel Marie. "A Potterhead's Progress: A Quest for Authenticity at the Wizarding World of Harry Potter." In *Playing Harry Potter: Essays and Interviews on Fandom and Performance*, edited by Lisa S. Brenner, 24–47. Jefferson: McFarland, 2015.

Graburn, Nelson H. "The Anthropology of Tourism." *Annals of Tourism Research* 10, no. 1 (1983): 9–33.

Grady, Maura, and Tony Magistrale. *The Shawshank Experience: Tracking the History of the World's Favorite Movie*. New York: Palgrave Macmillan, 2016.

Graml, Gundolf. "(Re)Mapping the Nation: *Sound of Music* Tourism and National Identity in Austria, ca 2000 CE." *Tourist Studies* 4, no. 2 (2004): 137–159.

Grau, Oliver. *Virtual Art: From Illusion to Immersion*. Translated by Gloria Custance. Cambridge: MIT Press, 2003.

Gray, Jonathan. *Show Sold Separately: Promos, Spoilers, and Other Media Paratexts*. New York: New York University Press. 2010.

Grossberg, Lawrence. "Is There a Fan in the House? The Affective Sensibility of Fandom." In *The Adoring Audience: Fan Culture and Popular Media*, edited by Lisa A. Lewis, 50–65. London: Routledge, 1992.

Gunelius, Susan. *Harry Potter: The Story of a Global Business Phenomenon*. Houndmills: Palgrave Macmillan, 2008.

Gwenllian-Jones, Sara. "Web Wars: Resistance, Online Fandom, and Studio Censorship." In *Quality Popular Television: Cult TV, the Industry and Fans*, edited by Mark Jancovich and James Lyons, 163–177. London: British Film Institute, 2003.

Halfacree, Keith. "'A Solid Partner in a Fluid World' and/or 'Line of Flight'? Interpreting Second Homes in the Era of Mobilities." *Norsk Geografisk Tidsskrift–Norwegian Journal of Geography* 65, no. 3 (2011): 144–153.

Hanna, Erin. "Be Selling You: *The Prisoner* as Cult and Commodity." *Television & New Media* 15, no. 5 (2014): 433–448.

Haring, Bruce. "'Friends' 25th Anniversary Spawns a Pop-Up Experience Straight Outta 1994." *Deadline*, July 29, 2019. https://deadline.com/2019/07/friends-pop-up-experience-tickets-1202655212/.

Harrington, C. Lee, and Denise D. Bielby. 2010. "Autobiographical Reasoning in Long-Term Fandom." *Transformative Works and Cultures* 5 (2010). https://doi.org/10.3983/twc.2010.0209.

———. "A Life Course Perspective on Fandom." *International Journal of Cultural Studies* 13, no. 5 (2010): 429–450.

Harrington, C. Lee, Denise D. Bielby, and Anthony R. Bardo. "Life Course

Transitions and the Future of Fandom." *International Journal of Cultural Studies* 14, no. 6 (2011): 567–590.

Harvey-Jenner, Catriona, and Dusty Baxter-Wright. "Hooray, FRIENDSFEST Is Back for a 12 Week Tour of the UK." *Cosmopolitan*, July 20, 2017. https://www.cosmopolitan.com/uk/entertainment/news/a44301/friends-fest-touring-round-country/.

Hellekson, Karen. "A Fannish Field of Value: Online Fan Gift Culture." *Cinema Journal* 48, no. 4 (2009): 113–118.

Hennig, Christoph. "Tourism: Enacting Modern Myths." Translated by Alison Brown. In *The Tourist as a Metaphor of the Social World*, edited by Graham Dann, 169–187. Oxon: CABI Publishing, 2002.

Heward, Emily. "FriendsFest Is Coming to Manchester This Weekend—with More Room Sets and THAT Pivot Scene." *Manchester Evening News*, July 30, 2018. https://www.manchestereveningnews.co.uk/whats-on/whats-on-news/friends-fest-manchester-tickets-heaton-14968617.

Hills, Matt. *"Doctor Who": The Unfolding Event—Marketing, Merchandising and Mediatizing a Brand Anniversary.* Basingstoke: Palgrave Macmillan, 2015.

———. *Fan Cultures.* London: Routledge, 2002.

———. "Torchwood's Trans-transmedia: Media Tie-ins and Brand 'Fanagement.'" *Participations* 9, no. 2 (2012): 409–428.

Hoebink, Dorus, Stijn Reijnders, and Abby Waysdorf. "Exhibiting Fandom: A Museological Perspective." *Transformative Works and Cultures* 16 (2014). https://doi.org/10.3983/twc.2014.0529.

Huhtamo, Erkki. "Encapsulated Bodies in Motion: Simulators and the Quest for Total Immersion." In *Critical Issues in Electronic Media*, edited by Simon Penny, 159–186. Albany: State University of New York Press, 1995.

Jancovich, Mark. "Cult Fictions: Cult Movies, Subcultural Capital and the Production of Cultural Distinctions." *Cultural Studies* 16, no. 2 (2002): 306–322.

Jansson, André. "A Sense of Tourism: New Media and the Dialectic of Encapsulation/Decapsulation." *Tourist Studies* 7, no. 1 (2007): 5–24.

———. "Spatial Phantasmagoria: The Mediatization of Tourism Experience." *European Journal of Communication* 17, no. 4 (2002): 429–443.

Jenkins, Henry. *Convergence Culture: Where Old and New Media Collide.* New York: New York University Press, 2006.

———. "The Future of Fandom." In *Fandom: Identities and Communities in a Mediated World*, edited by C. Lee Harrington, Jonathan Gray, and Cornel Sandvoss, 357–364. New York: New York University Press, 2007.

———. *Textual Poachers: Television Fans and Participatory Cultures.* New York: Routledge, 1992.

Jenkins, Olivia. "Photography and Travel Brochures: The Circle of Representation." *Tourism Geographies* 5, no. 3 (2003): 305–328.

Jenson, Joli. "Fandom as Pathology: The Consequences of Characterization." In *The Adoring Audience: Fan Culture and Popular Media*, edited by Lisa A. Lewis, 9–29. London: Routledge, 1992.

Johnson, Catherine. "Tele♥branding in TVIII: The Network as Brand and the Programme as Brand." *New Review of Film and Television Studies* 5, no. 1 (2007): 5–24.

———. *Telefantasy*. London: British Film Institute, 2005.

Jones, Bethan. "'*The Walking Dead* Family Is a Real Thing, Not Just a Hashtag': Theorizing Dissonance through Fan Tourism for *The Walking Dead* in Woodbury, Atlanta." *JOMEC Journal* 14 (2019): 53–70.

Joyce, Stephen. "Media Tourism and Conflict Heritage in Dubrovnik, Westeros." *The Journal of Popular Culture* 52, no. 6 (2019): 1387–1407.

Karpovich, Angelina I. "Theoretical Approaches to Film-Motivated Tourism." *Tourism and Hospitality Planning & Development* 7, no. 1 (2010): 7–20.

Keidl, Philipp Dominik. "Behind-the-(Museum)-Scenes: Fan-Curated Exhibitions as Tourist Attractions." In *The Routledge Companion to Media and Tourism*, edited by Maria Månsson, Annæ Buchmann, Cecilia Cassinger, and Lena Eskilsson, 297–306. London: Routledge, 2021.

Kim, Sangkyun. "Audience Involvement and Film Tourism Experiences: Emotional Places, Emotional Experiences." *Tourism Management* 33 (2012): 387–396.

———. "Extraordinary Experience: Re-enacting and Photographing at Screen Tourism Locations." *Tourism and Hospitality Planning & Development* 7, no. 1 (2010): 59–75.

Klein, Norman. "Electronic Baroque: Jerde Cities." In *You Are Here: The Jerde Partnership International*, edited by Ray Bradbury. London: Phaidon, 1999. http://artefact.mi2.hr/_a04/lang_en/theory_klein_en.htm.

Knox, Simone, and Kai Hanno Schwind. *Friends: A Reading of the Sitcom*. Cham: Palgrave Macmillan, 2019.

Koren-kuik, Meyrav. "Desiring the Tangible: Disneyland, Fandom, and Spatial Immersion." In *Fan CULTure: Essays on Participatory Fandom in the 21st Century*, edited by Kristin M. Barton and Jonathan Malcolm Lampley, 146–158. Jefferson: McFarland, 2014.

Kutulas, Judy. "Anatomy of a Hit: *Friends* and Its Sitcom Legacies." *The Journal of Popular Culture* 51, no. 5 (2018): 1172–1189.

Langley, Roger. *Portmeirion on Film and Television—The Full 70 Year Screen History*. Available from the shops in Portmeirion, 2011.

Larsen, Jonas. "Families Seen Sightseeing: Performativity of Tourist Photography." *Space and Culture* 8, no. 4 (2005): 416–434.

Larsen, Katherine. "(Re)Claiming Harry Potter Fan Pilgrimage Sites." In *Playing Harry Potter: Essays and Interviews on Fandom and Performance*, edited by Lisa S. Brenner, 38–54. Jefferson: McFarland, 2015.

Larsen, Katherine, and Lynn Zubernis. *Fandom at the Crossroads: Celebration, Shame and Fan/Producer Relationships*. Newcastle upon Tyne: Cambridge Scholars Publishing, 2012.

Leach, Joan. "Rhetorical Analysis." In *Qualitative Researching with Text, Image and Sound: A Practical Handbook*, edited by Martin W. Bauer and George Gaskell, 207–226. London: Sage, 2000.

Lee, Christina. "'Have Magic, Will Travel': Tourism and Harry Potter's United (Magical) Kingdom." *Tourist Studies* 12, no. 1 (2012): 52–69.

Lefferts, Brooke. "'Friends' Pop-Up Lets Sitcom's Fans Explore Show's Key Props." *Associated Press*, September 6, 2019. https://apnews.com/ceaf5e17 9abe494894eb3fe90551eb5e.

Leonard, Marion. "Exhibiting Popular Music: Museum Audiences, Inclusion and Social History." *Journal of New Music Research* 39, no. 2 (2010): 171–181.

Linden, Henrik, and Sara Linden. *Fans and Fan Cultures: Tourism, Consumerism and Social Media*. London: Palgrave Macmillan, 2017.

Lizardi, Ryan. *Mediated Nostalgia: Individual Memory and Contemporary Mass Media*. Lanham: Lexington Books, 2015.

Loughrey, Clarisse. "Friends: FriendsFest Is Back and Touring the UK This Time." *The Independent*, June 28, 2016. https://www.independent.co.uk/arts-enter tainment/tv/news/friends-friendsfest-is-back-and-touring-the-uk-this-time -a7106981.html.

Low, Setha M., and Irving Altman. "Place Attachment: A Conceptual Inquiry." In *Place Attachment*, edited by Irving Altman and Setha M. Low, 1–12. New York: Plenum Press, 1992.

Lucas, Damien. "Legendary TV Show Friends Takes Over Milton Keynes This Week for 25th Anniversary Summer Tour." *Milton Keynes Citizen*, September 4, 2019. https://www.miltonkeynes.co.uk/whats-on/things-to-do /legendary-tv-show-friends-takes-over-milton-keynes-week-25th-anniver sary-summer-tour-944459.

Lukas, Scott A. "Theming as a Sensory Phenomenon: Discovering the Senses on the Las Vegas Strip." In *The Themed Space: Locating Culture, Nation, and Self*, edited by Scott Lukas, 75–95. Plymouth: Lexington, 2007.

Lundberg, Christine, and Vassilios Ziakas, eds. *The Routledge Handbook of Popular Culture and Tourism*. London: Routledge, 2019.

Lundberg, Christine, Vassilios Ziakas, and Nigel Morgan. "Conceptualising On-Screen Tourism Destination Development." *Tourist Studies* 18, no. 1 (2018): 83–104.

MacCannell, Dean. *The Ethics of Sightseeing*. Berkeley: University of California Press, 2011.

———. *The Tourist: A New Theory of the Leisure Class*. 2nd ed. Berkeley: University of California Press, 1999. Originally published 1976.

Macionis, Niki, and Beverley Sparks. "Film-Induced Tourism: An Incidental

Experience." *Tourism Review International* 13, no. 2 (2009): 93–101.

Månsson, Maria. "Mediatized Tourism." *Annals of Tourism Research* 38, no. 4 (2011): 1634–1652.

Månsson, Maria, Annæ Buchmann, Cecilia Cassinger, and Lena Eskilsson, eds. *The Routledge Companion to Media and Tourism*. London: Routledge, 2021.

Martens, Todd. "Commentary: What Works, What's Missing and What Needs Fixing at Disney's Galaxy's Edge." *Los Angeles Times*, October 7, 2019. https://www.latimes.com/entertainment-arts/story/2019-10-07/disney land-star-wars-galaxys-edge-progress-report.

Massey, Doreen. *Place, Space, and Gender*. Cambridge: Polity, 1994.

Mathijs, Ernest, and Martin Barker. "Seeing the Promised Land from Afar: The Perception of New Zealand by Overseas *The Lord of the Rings* Audiences." In *How We Became Middle-earth: A Collection of Essays on "The Lord of the Rings,"* edited by Adam Lam and Natalia Oryshchuk, 107–128. Switzerland: Walking Tree Publishers, 2007.

McGinn, Colin. *Mindsight: Image, Dream, Meaning*. Cambridge: Harvard University Press, 2004.

McKee, Alan. "The Fans of Cultural Theory." In *Fandom: Identities and Communities in a Mediated World*, edited by C. Lee Harrington, Jonathan Gray, and Cornel Sandvoss, 88–97. New York: New York University Press, 2007.

Mitrasinovic, Miodrag. *Total Landscape, Theme Parks, and Public Space*. Aldershot: Ashgate, 2006.

Mittell, Jason. "Strategies of Storytelling on Transmedia Television." In *Storyworlds Across Media: Towards a Media-Conscious Narratology*, edited by Marie-Laure Ryan and Jan-Noël Thon, 253–277. Lincoln: University of Nebraska Press, 2014.

Mittermeier, Sabrina. "(Un)Conventional Voyages?—*Star Trek: The Cruise* and the Themed Cruise Experience." *The Journal of Popular Culture* 52, no. 6 (2019): 1372–1386.

Moores, Sean. *Media, Place, and Mobility*. Basingstoke: Palgrave Macmillan, 2012.

Mordue, Tom. "Performing and Directing Resident/Tourist Cultures in Heartbeat Country." *Tourist Studies* 1, no. 3 (2001): 233–252.

Mortensen, Christian Hviid. "Commemorating Popular Media Heritage: From Shrines of Fandom to Sites of Memory." In *The Routledge Companion to Media and Tourism*, edited by Maria Månsson, Annæ Buchmann, Cecilia Cassinger, and Lena Eskilsson, 267–276. London: Routledge, 2021.

Murray, Janet H. *Hamlet on the Holodeck: The Future of Narrative in Cyberspace*. New York: The Free Press, 1997.

Niallgirlalmighty. Tumblr post, July 23, 2014. http://niallgirlalmighty.tumblr .com/post/9268173743. Accessed October 12, 2015.

O'Brien, Danielle. "COULD WE BE MORE EXCITED? FriendsFest Is Returning to the UK . . . Here's What to Expect and How You Can Get Tickets NOW." *The Sun*, March 28, 2018. https://www.thesun.co.uk/fabulous/5918782 /friendsfest-is-returning-to-the-uk-heres-what-to-expect-and-how-you-can -get-tickets/.

Parker, Lilliana. "Editorial: Why Star Tours Is a Better Star Wars Attraction Than Millennium Falcon." *MiceChat*, December 19, 2019. https://www.micechat .com/238932-editorial-why-star-tours-is-a-better-star-wars-attraction-than -millennium-falcon/

Pearson, Roberta. "Bachies, Bardies, Trekkies, and Sherlockians." In *Fandom: Identities and Communities in a Mediated World*, edited by C. Lee Harrington, Jonathan Gray, and Cornel Sandvoss, 98–109. New York: New York University Press, 2007.

Peaslee, Robert Moses. "One Ring, Many Circles: The Hobbiton Tour Experience and a Spatial Approach to Media Power." *Tourist Studies* 11, no. 1 (2010): 37–53.

Peaslee, Robert Moses, and Rosalynn Vasquez. "*Game of Thrones*, Tourism, and the Ethics of Adaptation." *Adaptation* (2020): apaa012. https://doi.org/10 .1093/adaptation/apaa012.

People. "Explore the Instagram Worthy, Sold-Out *Friends* 25th Anniversary Pop-Up!" September 19, 2019. https://people.com/tv/explore-the-insta gram-worthy-sold-out-friends-25th-anniversary-pop-up/.

Pine, Joseph B., and James H. Gilmore. *The Experience Economy: Work Is Theatre & Every Business a Stage*. Boston: Harvard Business Press, 1999.

Porter, Jennifer E. "Pilgrimage and the IDIC Ethic: Exploring *Star Trek* Convention Attendance as Pilgrimage." In *Intersecting Journeys: The Anthropology of Pilgrimage and Tourism*, edited by Ellen Badone and Sharon R. Roseman, 160–179. Urbana: University of Illinois Press, 2004.

Pritchard, Julia. "Oh, My, God! Lydia Bright Flashes Some Leg in Floral Maxi as She Fangirls over Childhood Favourite Maggie Wheeler (aka Janice) at Friendsfest." *MailOnline*, August 24, 2016. https://www.dailymail.co.uk /tvshowbiz/article-3755619/Lydia-Bright-flashes-leg-floral-maxi-fangirls -childhood-favourite-Maggie-Wheeler-aka-Janice-Friendsfest.html.

Rakić, Tijana, and Donna Chambers. "Rethinking the Consumption of Places." *Annals of Tourism Research* 39, no. 3 (2012): 1612–1633.

Reijnders, Stijn. *Places of the Imagination: Media, Tourism, Culture*. Farnham: Ashgate Publishing, 2011.

Roberson, Richard, and Maura Grady. "The 'Shawshank Trail': A Cross Disciplinary Study in Film Induced Tourism and Fan Culture." *AlmaTourism* 6, no. 4 (2015): 47–66.

Rodaway, Paul. *Sensuous Geographies: Body, Sense, and Place*. New York: Routledge, 2002.

Rodman, Gilbert. *Elvis after Elvis: The Posthumous Career of a Living Legend*. London: Routledge, 1996.

Roesch, Stefan. *The Experiences of Film Location Tourists*. Bristol: Channel View Publications, 2009.

Rojek, Chris. "Indexing, Dragging, and the Social Construction of Tourist Sites." In *Touring Cultures: Transformations of Travel and Theory*, edited by Chris Rojek and John Urry, 52–74. London: Routledge, 1997.

Ryan, Marie-Laure. *Narrative as Virtual Reality: Immersion and Interactivity in Literature and Electronic Media*. Baltimore: Johns Hopkins University Press, 2001.

Sadler, William J., and Ekaterina V. Haskins. "Metonymy and the Metropolis: Television Show Settings and the Image of New York City." *Journal of Communication Inquiry* 29, no. 3 (2005): 195–216.

Saler, Michael. *As-If: Modern Enchantment and the Literary Prehistory of Virtual Reality*. New York: Oxford University Press, 2012.

Sandvoss, Cornel. *Fans: The Mirror of Consumption*. Cambridge: Polity, 2005.

———. "'I♥Ibiza': Music, Place, and Belonging." In *Popular Music Fandom: Identities, Roles, and Practices*, edited by Mark Duffet, 115–145. New York: Routledge, 2014.

Sarner, Lauren. "Overhyped 'Friends' Pop-Up Is a 25th Anniversary Rip-Off." *New York Post*, September 6, 2019. https://nypost.com/2019/09/06/over hyped-friends-pop-up-is-a-25th-anniversary-rip-off/.

Scannell, Leila, and Robert Gifford. "Defining Place Attachment: A Tripartite Organizing Framework." *Journal of Environmental Psychology* 30, no. 1 (2010): 1–10.

Seamon, David. *A Geography of the Lifeworld: Movement, Rest, and Encounter*. London: Croom Helm, 1979.

Selling, Kim. "Fantastic Neomedievalism: The Image of the Middle Ages in Popular Fantasy." In *Flashes of the Fantastic: Selected Papers from "The War of the Worlds" Centennial, Nineteenth International Conference on the Fantastic in the Arts*, edited by David Ketterer, 211–218. Westport: Praeger, 2004.

Selvam, Ashok. "Look Around the 'Saved by the Bell' Pop-Up, Debuting Today in Chicago." *Eater Chicago*, June 1, 2016. https://chicago.eater.com/2016/6/1 /11827664/saved-by-the-bell-photos-gallery-menus-chicago.

Short, Sue. *Cult Telefantasy Series: A Critical Analysis of "The Prisoner," "Twin Peaks," "The X-Files," "Buffy the Vampire Slayer," "Lost," "Heroes," "Doctor Who" and "Star Trek."* Jefferson: McFarland, 2011.

Silvestri, Lisa Ellen. "Memeingful Memories and the Art of Resistance." *New Media & Society* 20, no. 11 (2018): 3997–4016.

Sorkin, Michael. "See You in Disneyland." In *Variations on a Theme Park: The New American City and the End of Public Space*, edited by Michael Sorkin, 205–249. New York: Hill and Wang, 1992.

Stanfill, Mel. *Exploiting Fandom: How the Media Industry Seeks to Manipulate Fans.* Iowa City: University of Iowa Press, 2019.

———. "'They're Losers, but I Know Better': Intra-Fandom Stereotyping and the Normalization of the Fan Subject." *Critical Studies in Media Communication* 30, no. 2 (2013): 117–134.

Sternbergh, Adam. "Is *Friends* Still the Most Popular Show on TV?" *Vulture,* March 21, 2016. https://www.vulture.com/2016/03/20-somethings -streaming-friends-c-v-r.html.

Strause, Jackie. "'Friends' Turns 25: Inside the NBC Sitcom's Pop-Up Return to New York." *The Hollywood Reporter*, September 6, 2019. https://www.holly woodreporter.com/live-feed/friends-turns-25-inside-new-york-city-pop-up -1237174.

Sullivan, C. W., III. "High Fantasy." In *International Companion Encyclopedia of Children's Literature*, edited by Peter Hunt, 436–446. London: Routledge, 2004.

Sundbo, Jon, and Flemming Sørenson. "Introduction to the Experience Economy." In *Handbook on the Experience Economy*, edited by Jon Sundbo and Flemming Sørenson, 1–17. Cheltenham: Edward Elgar, 2013.

Sylt, Christian. "Revealed: The World's Fastest-Growing Theme Park." *Forbes,* June 30, 2019. https://www.forbes.com/sites/csylt/2019/06/30/revealed -the-worlds-fastest-growing-theme-park/.

Thompson, Lauren Jade. "'It's Like a Guy Never Lived Here!': Reading the Gendered Domestic Spaces of *Friends*." *Television & New Media* 19, no. 8 (2018): 758–774.

Thrift, Nigel. *Spatial Formations.* London: Sage, 1996.

Todd, Anne Marie. "Saying Goodbye to *Friends*: Situation Comedy as Lived Experience." *The Journal of Popular Culture* 44, no. 4 (2011): 855–872.

Torchin, Leshu. "Location, Location, Location: The Destination of the Manhattan TV Tour." *Tourist Studies* 2, no. 3 (2002): 247–266.

Tosenberger, Catherine. "Homosexuality at the Online Hogwarts: Harry Potter Slash Fanfiction." *Children's Literature* 36, no. 1 (2008): 185–207.

Tuan, Yi-Fu. *Space and Place: The Perspective of Experience.* Minneapolis: University of Minnesota Press, 1977.

Turchiano, Danielle. "'Friends' at 25: How Warner Bros. TV Built an Experiential Empire." *Variety,* September 19, 2019. https://variety.com/2019/tv/fea tures/friends-25th-anniversary-stunts-partnerships-popup-lego-apps-pot tery-barn-studio-tour-interview-1203331201/.

Turner, Kyle. "The Friends Pop-Up Captures Everything About the Show Except What Made It Great." *Slate,* September 13, 2019. https://slate.com/culture /2019/09/friends-tv-pop-up-experience-shop-new-york.html.

Turner, Victor. *The Ritual Process: Structure and Anti-Structure.* Ithaca: Cornell University Press, 1977.

Bibliography

Tzanelli, Rodanthi, and Majid Yar. "*Breaking Bad*, Making Good: Notes on a Tele-visual Tourist Industry." *Mobilities* 11, no. 2 (2016): 188–206.

Urry, John. *The Tourist Gaze*. 2nd ed. London: Sage, 2002.

Urry, John, and Jonas Larsen. *The Tourist Gaze 3.0*. London: Sage, 2011.

Waade, Anne Marit. "Mind the Gap: Interdisciplinary Approaches to Media and Tourism." In *The Routledge Companion to Media and Tourism*, edited by Maria Månsson, Annæ Buchmann, Cecilia Cassinger, and Lena Eskilsson, 20–26. London: Routledge, 2021.

Wang, Ning. "Rethinking Authenticity in Tourism Experience." *Annals of Tourism Research* 26, no. 2 (1999): 349–370.

Watson, Nicola. *The Literary Tourist: Readers and Places in Romantic & Victorian Britain*. Basingstoke: Palgrave Macmillan, 2006.

Waysdorf, Abby. "'I Don't Think the Two Would Be the Same without Each Other': Portmeirion as Unintentional Paratext." *JOMEC Journal* 14 (2019): 33–52.

Whitbrook, James. "*Star Wars: Galaxy's Edge* Is Almost Too Alien for Its Own Good." *io9*, November 27, 2019. https://io9.gizmodo.com/star-wars-galaxys-edge-is-almost-too-alien-for-its-own-1840053738.

Williams, Rebecca. "Funko Hannibal in Florence: Fan Tourism, Participatory Culture, and Paratextual Play." *JOMEC Journal* 14 (2019): 71–90.

———. *Post-Object Fandom: Television, Identity, and Self-Narrative*. New York: Bloomsbury, 2015.

———. *Theme Park Fandom: Spatial Transmedia, Materiality and Participatory Cultures*. Amsterdam: Amsterdam University Press, 2020.

———. "Theme Parks in the Time of the COVID-19 Pandemic." In *Pandemic Media: Preliminary Notes toward an Inventory*, edited by Philipp Dominik Keidl et al. Meson Press, 2020. https://pandemicmedia.meson.press/chapters/space-scale/theme-parks-in-the-time-of-the-covid-19-pandemic/.

Williams-Ellis, Clough. *Portmeirion: The Place and Its Meaning*. Rev. ed. Porthmadog: Portmeirion Shops, 2014.

Young, Helen. "Approaches to Medievalism: A Consideration of Taxonomy and Methodology through Fantasy Fiction." *Parergon* 27, no. 1 (2010): 163–179.

Zachry, Mark. "Rhetorical Analysis." In *The Handbook of Business Discourse*, edited by Francesca Bargiela-Chiappini, 68–79. Edinburgh: Edinburgh University Press, 2009.

Zubernis, Lynn, and Katherine Larsen. "Make Space for Us! Fandom in the Real World." In *A Companion to Media Fandom and Fan Studies*, edited by Paul Booth, 145–159. Hoboken: Wiley Blackwell, 2018.

Index

Fandom & Culture

READINGS IN LABOR RELATIONS

From
The Wall Street Journal

Karl O. Mann
Professor of Industrial Relations
Rider College

DOW JONES BOOKS — *ACADEMIC SERIES*
Princeton, New Jersey 08540

...k of readings prepared for the ...piler, as well as for related courses ...sity libraries. For information about ..te to:

DOW JONES BOOKS — Academic Series
P.O. Box 300
Princeton, N.J. 08540

Library of Congress Cataloging in Publication Data

Mann, Karl O.   comp.
   Readings in labor relations from The Wall Street
Journal.

   (Academic series)
   1. Trade-unions — United States — Addresses, essays,
lectures. 2. Collective bargaining — United States —
Addresses, essays, lectures. I. The Wall Street Journal.
II. Title.
HD6508.M265   331.88'0973   74-23668
ISBN 0-87128-850-8 pbk.